The Little Book of Probability

Second Edition[1]

Essentials of Probability and Statistics for Stochastic Processes and Simulation

Bruce Wayne Schmeiser
School of Industrial Engineering
Purdue University

and

Michael Robert Taaffe
Department of Industrial and Systems Engineering
Virginia Tech

---

[1] Revised January 30, 2012

The Little Book of Probability
Second Edition

© Bruce W. Schmeiser and Michael R. Taaffe
ISBN-13: 978-1452882925
ISBN-10: 1452882924

Table of Contents

Topic	Pages
Table of Contents	i
Preface	ii
Set-Theory Review	1 – 3
Probability Basics	5 – 14
Discrete Random Variables	15 – 31
Continuous Random Variables	33 – 49
Random Vectors and Joint Probability Distributions	51 – 70
Descriptive Statistics	71 – 74
Parameter Point Estimation	75 – 83
Greek Alphabet	85
A Few Mathematics Results	87 – 89

Preface

The Little Book of Probability contains the definitions and results from an undergraduate course in elementary probability and statistics (for example, IE 230 at Purdue, or STAT 4105 at Virginia Tech). The purpose of *The Little Book of Probability* is to provide a complete, clear, and concise compendium.

The purpose of course lectures, homework assignments, professors' office hours, and course textbook, is to help *understand* the meanings and implications of the results listed in *The Little Book of Probability* via discussion and examples.

The Little Book of Probability provides the language that is used in stochastic modeling. These essentials are prerequisites for undergraduate courses in stochastic processes, probability modeling and analysis, stochastic operations research, probabilistic operations research, and simulation modeling and analysis. Most instructors assume that *all* of the material in this book has been mastered before enrolling in these I(S)E courses as well as undergraduate I(S)E courses in statistical quality control, or any applied statistics courses such as experiment design, linear regression, etc.

Although the focus of *The Little Book of Probability* is on elementary, non-measure-theoretic probability results, for completeness we also include brief sections on elementary descriptive statistics and parameter estimation.

Finally we include a listing of the Greek alphabet and some elementary mathematical results for the readers' convenience.

Please report to us any errors – mathematical or typographical – via email address: taaffe@vt.edu. – January 30, 2012.

Set-Theory Review

A *set* is a collection of items;
  each item is called a *member* or *element* of the set.

Braces, { }, indicate sets.

If a set A has only members $x, y,$ and z, write $A = \{x, y, z\}$.

If x is an element of the set A, write $x \in A$.

If a set has members defined by a condition, write
  $A = \{x \,|\, x \text{ satisfies the condition}\}$.
  The vertical line is read "such that."

The largest set is the *universe*, the set containing all relevant items.

Denote the universal set by U.

The smallest set is the *empty set* (also called the *null set*), which contains no items. It is denoted by $\emptyset$ or, occasionally, by { }.

If all members of a set A are contained in a set B, then A is a *subset* of B, written $A \subseteq B$. Every set is a subset of U.

If two sets A and B contain the same members, then they are equal, written $A = B$.

The *complement* of a set A is the set of all items not contained in the set; that is, $A' = \{x \,|\, x \notin A\}$.
  The complement sometimes is written as A^c.

The *union* of two sets A and B is the set of items contained in one or both of the sets;
  that is, $A \cup B = \{x \,|\, x \in A \text{ or } x \in B\}$.

–1–

The *intersection* of two sets A and B is the set of items contained in both sets; that is,
$A \cap B = \{x \mid x \in A \text{ and } x \in B\}$. The intersection also is written as AB and $\{x \mid x \in A, x \in B\}$.

Sets A and B are *mutually exclusive* if $A \cap B = \emptyset$.

Sets $E_1, E_2, \ldots, E_n$ *partition* the set A if every member of A lies in exactly one of the n partitioning sets.

(Equivalently, if A is a subset of the union of the sets $E_1, E_2, \ldots, E_n$, and the sets $E_1, E_2, \ldots, E_n$ are mutually exclusive, then the sets $E_1, E_2, \ldots, E_n$ form a partition of A. Notationally, $A \subseteq \bigcup_{i=1}^{n} E_i$ and $E_i \cap E_j = \emptyset$ for every pair E_i and E_j.)

DeMorgan's Laws. For any sets A and B,
$$(A \cup B)' = A' \cap B'$$
and
$$(A \cap B)' = A' \cup B'.$$

Distributive Laws. For any sets A, B, and C,
$$(A \cup B) \cap C = (A \cap C) \cup (B \cap C)$$
and
$$(A \cap B) \cup C = (A \cup C) \cap (B \cup C).$$

The *cardinal number* (*cardinality*) of a set A, denoted by $\#(A)$, is the number of members in A.

- The set A is *finite* if $\#(A)$ is finite; otherwise A is *infinite*.

- The infinite set A is *countably infinite* if its members can be counted (to *count* a set means to assign a unique integer to each member); otherwise A is *uncountably infinite* (e.g., the set of real numbers).

Set exclusion. For any set $A = \{a_1, a_2, \ldots, a_n\}$, the exclusion of member a_i is indicated by $A \backslash a_i$.

The *open interval* (a, b) is the set $\{x \mid a < x < b\}$, all real numbers between, but not including a or b. Square brackets are used to include a and/or b. Except for $[a, a]$, the *closed interval* containing only the one number a, non-empty intervals are uncountably infinite.

The *real-number line* is the open interval $\mathbb{R} \equiv (-\infty, \infty)$.

$\mathbb{R}^+ = [0, \infty)$, the set of non-negative real numbers.

$\mathbb{R}^n$ is the set of n-dimensional real numbers.

A *function* assigns a single *value* to each *argument*. The set of arguments is called the *domain* and the set of values lies within the *range*. For example, $f(x) = x^2$ has domain $\mathbb{R}$ and range $\mathbb{R}^+$. For example, let the domain be a set of students and let the function be the student weight in pounds; then the range is the set of student weights.

Probability Basics

All of probability and statistics depends upon the concept of a *random experiment*. Every *replication* of the random experiment results in one *outcome* from the set of all possible outcomes.

A set containing all possible outcomes is called a *sample space*. A random experiment can have many sample spaces. The chosen sample space, denoted by Ω, must be sufficient to answer the questions at hand. For example, sample spaces typically associated with the random experiment of tossing a six-sided die include $\Omega = \{1, 2, 3, 4, 5, 6\}$ and $\Omega = \{even, odd\}$.

A set, say E, is an *event* if it is a subset of Ω; that is, if $E \subseteq \Omega$.

– The event E *occurs* if it contains the outcome; otherwise E *does not occur*.

– In practice, an event E can be a verbal statement that is true if the event occurs and false if the event does not occur.

The complement of the event E is the event E'.

– The statement for E' is the negation of the statement for E.

– E' occurs if (and only if) the outcome does not lie in E.

– Exactly one of E and E' occurs.

The union of events E_1 and E_2 is the event $E_1 \cup E_2$.

- Either or both of the statements for E_1 and E_2 must be true for $E_1 \cup E_2$ to occur.
- $E_1 \cup E_2$ occurs if (and only if) the outcome lies in E_1, or in E_2, or in both.
- The complement of $E_1 \cup E_2$ is $(E_1 \cup E_2)' = E'_1 \cap E'_2$. That is, $(E_1 \cup E_2)'$ occurs if E_1 does not occur and E_2 does not occur.

The intersection of events E_1 and E_2 is the event $E_1 \cap E_2$.

- The statements for E_1 and E_2 must both be true for $E_1 \cap E_2$ to occur.
- $E_1 \cap E_2$ occurs if (and only if) the outcome lies in both E_1 and E_2.
- The complement of $E_1 \cap E_2$ is $(E_1 \cap E_2)' = E'_1 \cup E'_2$. That is, $(E_1 \cap E_2)'$ occurs if either E_1 does not occur, or E_2 does not occur, or neither occurs.

An experiment always has multiple events.

- The largest event is Ω, which always occurs.
- The smallest event is the empty set, $\emptyset$, which never occurs.
- For every event E, $\emptyset \subseteq E \subseteq \Omega$

Two events, say E_1 and E_2, are *mutually exclusive* if they cannot both occur; that is, if $E_1 \cap E_2 = \emptyset$. More generally, n events, say $E_1, E_2, \ldots, E_n$, are *mutually exclusive* if only one can occur; that is, if $E_i \cap E_j = \emptyset$ for every pair of events.

The correspondence between set theory and probability theory is that the universe is the sample space, the elements are outcomes of the random experiment, and subsets of the sample space are events.

The *probability* of an event E, denoted by P(E), is a numerical measure of how likely the event E is to occur when the experiment is performed.

P($\cdot$) is a function.
- The argument of P($\cdot$) is an event.
- The value of P($\cdot$) is a real number in $[0, 1]$.
  For example, P(E_1) = .8, P($E_1 \cup (E_2 \cap E_3)$) = $1/\pi$, or P($\{a, b, c\}$) = .05.

There are two commonly used interpretations of probability.

Relative frequency: If the random experiment were replicated infinitely often, then P(E) is the fraction of the replications in which E occurs. (This interpretation is sometimes called "objective" or "frequentist." For example, the probability of a fair coin toss resulting a *head* is 0.5)

Subjective: P(E) is a measure of belief about the likelihood that E will occur in a particular replication. (This interpretation is sometimes called "Bayesian." For example the probability of a particular team winning a particular sports game might be estimated to be 0.3.)

Alternative statements of probability.

"The odds of E are 2 in 3" is equivalent to "P(E) = 2/3."

"The odds of E are 2 to 3" is equivalent to "P(E) = 2/5."

"E has a 70% chance" is equivalent to "P(E) = 0.7."

"E has a 50–50 chance" is equivalent to "P(E) = 0.5."

A baseball player batting 282 has hit successfully with relative frequency 0.282.

All results of probability follow from three *axioms*.

Axiom 1. $P(\Omega) = 1$.
(That is, the "sure" event has probability one.)

Axiom 2. For every event E, $0 \leq P(E)$.
(That is, probabilities are non-negative.)

Axiom 3. If E_1 and E_2 are mutually exclusive events, then
$$P(E_1 \cup E_2) = P(E_1) + P(E_2).$$

(That is, if two events cannot both occur, then the probability that one or the other occurs is the sum of their probabilities.)

The three axioms yield eight immediate results.

Result. (Complement) If E is an event, then $P(E') = 1 - P(E)$.

Result. (Null Event) $P(\emptyset) = 1 - P(\Omega) = 0$.
(The "impossible" event has probability zero.)

Result. (Dominance) If $E_1 \subset E_2$, then $P(E_1) \leq P(E_2)$.
(That is, if two events differ only in that one contains more outcomes than the other, the larger event cannot be less likely.)

Result. (Equally likely events) If equi-probable events $E_1, E_2, \ldots, E_n$ partition the sample space, then $P(E_i) = 1/n$ for $i = 1, 2, \ldots, n$.

Definition. Simple Sample Space: If Ω is a finite sample space with all outcomes ω_i equally likely, then $P(\{\omega_i\}) = 1/\#(\Omega)$, for every $i = 1, \ldots, \#(\Omega)$.

Result. The probability of an intersection of events is also commonly written as

$$P(A \cap B) \equiv P(A, B)$$

and more generally

$$P\left(\bigcap_{i=1}^{n} E_i\right) \equiv P(E_1, E_2, \ldots, E_n).$$

Result. (Principle of Inclusion-Exclusion). For any two events E_1 and E_2,

$$P(E_1 \cup E_2) = P(E_1) + P(E_2) - P(E_1 \cap E_2).$$

More generally, for any three events E_1, E_2, and E_3,

$$\begin{aligned} P(E_1 \cup E_2 \cup E_3) &= P(E_1) + P(E_2) + P(E_3) \\ &- P(E_1 \cap E_2) - P(E_1 \cap E_3) \\ &- P(E_2 \cap E_3) \\ &+ P(E_1 \cap E_2 \cap E_3). \end{aligned}$$

Yet more generally, for n events continue to alternate signs; i.e., if $E_1, E_2, \ldots, E_n$ are any n events, then

$$P\left(\bigcup_{i=1}^{n} E_i\right) = \sum_{i=1}^{n} P(E_i)$$
$$- \sum_{i=1}^{n-1} \sum_{j=i+1}^{n} P(E_i \cap E_j)$$
$$+ \sum_{i=1}^{n-2} \sum_{j=i+1}^{n-1} \sum_{k=j+1}^{n} P(E_i \cap E_j \cap E_k)$$
$$-$$
$$\vdots$$
$$+ (-1)^{n+1} P\left(\bigcap_{i=1}^{n} E_i\right)$$

A sequence of decreasing upper bounds on the probability is formed by truncating the right-hand side after positive terms and a sequence of increasing lower bounds on the probability is formed by truncating the right-hand side after negative terms.

Result. (Unions) If events $E_1, E_2, \ldots, E_n$ are mutually exclusive, then

$$P\left(\bigcup_{i=1}^{n} E_i\right) = \sum_{i=1}^{n} P(E_i).$$

(That is, if only one of the n events can occur, then the probability that one of them does occur is the sum of their probabilities. The usual proof for finite n is by induction using Axiom 3. Proving the result for infinite n is advanced.)

Result. (Boole's Inequality) For events $E_1, E_2, \ldots, E_n$,
$$P\left(\bigcup_{i=1}^{n} E_i\right) \leq \sum_{i=1}^{n} P(E_i).$$

Definition. The *conditional probability* of an event E_1, given that an event B has occurred, is
$$P(E_1|B) = \frac{P(E_1 \cap B)}{P(B)}.$$

– If $P(B) = 0$, then $P(E_1|B)$ is undefined.
– The *given* event B is assumed to have occurred; that is, the outcome is in B.
– The given event B becomes the (shrunken) sample space.
– $P(E_1) \equiv P(E_1|\Omega)$ is the *unconditional* or *marginal* probability of E.

Definition. *Multiplication Rule.* If $P(E_1) > 0$ and $P(B) > 0$, then
$$P(E_1 \cap B) = P(E_1|B) P(B) = P(B|E_1) P(E_1);$$
otherwise $P(E_1 \cap B) = 0$.

Comment: $P(E_1|B) P(B) = P(E_1 \cap B)$ is sometimes called *unconditioning*.

Definition. *Baby Bayes Rule.* If $P(B) > 0$, and $P(E_1) > 0$, then
$$P(E_1|B) = \frac{P(B|E_1)P(E_1)}{P(B)}.$$

Definition. (The full) *Bayes Rule* is obtained from the Baby Bayes Rule by expanding $P(B)$ using the Total Probability Theorem.

Definition. *Total Probability Theorem.* (Applying this theorem is sometimes called *conditioning*).
For any events B and E_1,

$$\begin{aligned} P(B) &= P(B \cap E_1) + P(B \cap E_1') \\ &= P(B|E_1)P(E_1) + P(B|E_1')P(E_1'). \end{aligned}$$

More generally, if events $E_1, E_2, \ldots, E_n$ partition Ω, then for any event B

$$P(B) = \sum_{i=1}^{n} P(B \cap E_i) = \sum_{i=1}^{n} P(B|E_i)P(E_i).$$

Definition. *Extended Multiplication Rule.* If $P(A_1) > 0, P(A_2) > 0, \ldots, P(A_n) > 0$, then

$$\begin{aligned} P(A_1 \cap A_2 \cap \cdots \cap A_n) &= P(A_1)P(A_2|A_1)P(A_3|A_1 \cap A_2) \\ &\quad \cdots P(A_n|A_1 \cap \cdots \cap A_{n-1}). \end{aligned}$$

Result. (Independent events) The following four statements are equivalent; that is, either all are false or all are true.

1. Events A and B are independent.
2. $P(A \cap B) = P(A)P(B)$.
3. $P(A|B) = P(A)$.
4. $P(B|A) = P(B)$.

(Statement 2 is often chosen to define event independence.)

Result. (Independence of complements) The following four statements are equivalent.

1. Events A and B are independent.
2. Events A' and B are independent.
3. Events A and B' are independent.
4. Events A' and B' are independent.

Result. If events A and B are non-empty subsets of a sample space Ω and are mutually exclusive, then A and B are dependent.

Result. If events A and B are not mutually exclusive, then A and B may be dependent or independent.

Definition. The n events $A_1, A_2, \ldots, A_n$ are (*jointly*) *independent* if and only if for $i, j = 1, \ldots, n$ where $i \neq j$,

$$P(A_i \cap A_j) = P(A_i)P(A_j),$$

and for $i, j, k = 1, \ldots, n$ where $i \neq j, i \neq k, j \neq k$,

$$P(A_i \cap A_j \cap A_k) = P(A_i)P(A_j)P(A_k), \text{ and}$$

..., for $k = 1, \ldots, n$,

$$P\left(\bigcap_{\substack{i=1 \\ i \neq k}}^{n} A_i\right) = \prod_{\substack{i=1 \\ i \neq k}}^{n} P(A_i)$$

and

$$P\left(\bigcap_{i=1}^{n} A_i\right) = \prod_{i=1}^{n} P(A_i).$$

Definition. *Pairwise independence* requires only that every pair of events be independent.

Result. Joint independence implies pairwise independence; pairwise independence does not imply joint independence. Two examples:

– Consider the random experiment of tossing a fair coin three times.

$$\begin{aligned}
\text{Let } A &= \text{"3 or 2 heads,"} \\
&= \{(h,h,h),(h,h,t),(h,t,h),(t,h,h)\}, \\
B &= \text{"3 or 1 heads,"} \\
&= \{(h,h,h),(h,t,t),(t,h,t),(t,t,h)\}, \\
\text{and } C &= \text{"last flip is a head"} \\
&= \{(h,h,h),(t,h,h),(h,t,h),(t,t,h)\}.
\end{aligned}$$

Observe that
- $P(A \cap B \cap C) = 1/8 = P(A)P(B)P(C)$, but (A,B,C) are jointly dependent because
- (A,B) are pairwise dependent, with $P(A \cap B) = 1/8 \neq P(A)P(A) = (1/2)^2$.

– Consider the random experiemt of tossing a fair coin two times.

$$\begin{aligned}
\text{Let } A &= \text{"head on the first flip"} = \{(h,h),(h,t)\}, \\
B &= \text{"head on the second flip"} = \{(h,h),(t,h)\}, \\
\text{and } C &= \text{"exactly one head"} = \{(h,t),(t,h)\}.
\end{aligned}$$

Observe that
- $P(A \cap B \cap C) = 0 \neq P(A)P(B)P(C) = (1/2)^3$; thus (A,B,C) are dependent, even though
- $(A,B),(A,C)$ and (B,C) are each pairwise independent.

Discrete Random Variables and Probability Distributions

Random variables are a convenient way to define large families of events.

Definition. A *random variable* is a function that assigns a real number to each outcome in the sample space of a random experiment; i.e.,

$$X(\omega_i) = x_i,$$

where $X(\cdot)$ is a random variable and for every $\omega_i \in \Omega$, there is a unique real number x_i.

- The domain of $X(\cdot)$ is Ω.
- The range of $X(\cdot)$ is a subset of $\mathbb{R}$.
- A random variable is a deterministic function and is *neither* random nor variable.
- Traditionally, random variables are denoted by upper-case letters toward the end of the English alphabet; e.g., X, Y.
- In practice, a random variable can be a verbal statement; e.g., the experiment is to select a random student and $X(\omega_i) =$ "the student i's grade point average."
- Events can be constructed from random variables (we say that an event (a subset of the sample space) is induced by (or defined by) a random variable taking on a value); e.g., $X > 3$ defines (or induces) an event (defined by the subset of students whose grade-point average is greater than 3.0). So we interpret the shorthand notation

event $X > 3 \iff$ event $\{\omega | X(\omega) > 3\}$. This leads to commonly used shorthand notation, such as

$$P(X = x) \text{ rather than } P(\{\omega | X(\omega) = x\})$$

and

$$P(X \leq x) \text{ rather than } P(\{\omega | X(\omega) \leq x\}),$$

where $\omega \in \Omega$ and $x \in \mathbb{R}$.

Definition. The *probability distribution* of a random variable $X(\cdot)$ is a description (in whatever form) of the likelihoods associated with all values in $\mathbb{R}$.

Definition. The *cumulative distribution function*, often abbreviated *cdf*, of a random variable $X(\cdot)$ is

$$\begin{aligned} F(x) &\equiv F_X(x) \\ &\equiv P(X \leq x) \\ &\equiv P(X(\omega) \leq x) \\ &\equiv P(\{\omega | X(\omega) \leq x\}) \end{aligned}$$

for $x \in \mathbb{R}$ and $\omega \in \Omega$.

Result. For every random variable $X(\cdot)$, if $a \leq b$, then

$$\begin{aligned} P(a < X \leq b) &\equiv P(\{\omega | a < X(\omega) \leq b\}) \\ &= F(b) - F(a), \end{aligned}$$

where $\omega \in \Omega$ and $a, b \in \mathbb{R}$.

Result. Every cdf is a monotone nondecreasing,
right-continuous function whose left and right limits are
0, and 1, respectively; i.e.,

– For $x \leq y$, $F(x) \leq F(y)$
(This is a special case of $P(A) \leq P(B)$ if $A \subseteq B$.)
– $\lim_{x \to -\infty} F(x) = 0$,
– $\lim_{x \to \infty} F(x) = 1$.

Definition. A random variable is *discrete* if its range is finite or countably infinite subset of $\mathbb{R}$.

Definition. A random variable is *continuous* if its range is uncountably infinite subset of $\mathbb{R}$.

Definition. A random variable is *mixed* if its range is the union of both countable and uncountable subsets of $\mathbb{R}$.

– Often, discrete random variables arise from *counting*.
– Often, continuous random variables arise from *measuring*.

Definition. For a discrete random variable $X(\cdot)$, the *probability mass function*, often abbreviated *pmf*, is $f(x) \equiv f_X(x) \equiv P(X = x) \equiv P(\{\omega | X(\omega) = x\})$ for $x \in \mathbb{R}$ and $\omega \in \Omega$.

Result. For a discrete random variable $X(\cdot)$ the cdf is

$$\begin{aligned}
F_X(x) &\equiv F(x) \\
&\equiv \sum_{\text{all } x_i \leq x} f(x_i) \\
&\equiv \sum_{\text{all } x_i \leq x} P(\{\omega | X(\omega) = x_i\}) \\
&\equiv P\left(\bigcup_{\text{all } x_i \leq x} \{\omega | X(\omega) = x_i\} \right) \\
&\equiv P(\{\omega | X(\omega) \leq x\})
\end{aligned}$$

for all $x \in \mathbb{R}$.

Definition. For a discrete random variable $X(\cdot)$ the *support* of $f(\cdot)$ is the set of real-numbers x_i, $i \in \mathbf{I}$, where $\mathbf{I}$ is the finite or countably infinite index set. So

$$f(x) = \begin{cases} f(x_i), & \text{for} \quad x = x_i, i \in \mathbf{I} \\ 0, & \text{otherwise} \end{cases}$$

Result. $f(x_i) = F(x_i) - F(x_{i-1})$.

Definition. The *mean* (or *first moment* or *first raw moment*, or *first power moment*, or the *expectation*) of a random variable $X(\cdot)$ (discrete, continuous, or mixed) is the constant

$$\begin{aligned}
E[X] &\equiv -\int_{-\infty}^{0} P(X \leq x)\, dx + \int_{0}^{\infty} P(X > x)\, dx \\
&= -\int_{-\infty}^{0} F(x)\, dx + \int_{0}^{\infty} (1 - F(x))\, dx
\end{aligned}$$

Comment. We refer to the above definition as the 'tail-probability definition" of the mean.

Definition. For a discrete random variable $X(\cdot)$, the mean is the constant $\mathrm{E}[X] \equiv \sum_{\text{all } x_i} x_i f(x_i)$.

Comment.

- The mean is also commonly denoted by μ or, more explicitly, by μ_X.
- Because $\sum_{\text{all } x_i} f(x_i) = 1$, the mean is the *first moment*, or *center of gravity*.

Comment. $\mathrm{E}[X]$ is a constant that lies between the smallest and the largest possible values of the random variable $X(\cdot)$.

Comment. $\mathrm{E}[X]$ does not need to be equal to one of the possible values of the random variable $X(\cdot)$ and should not be rounded to a possible value.

Comment. The units of $\mathrm{E}[X]$ are those of the random variable $X(\cdot)$.

Comment. The cdf is denoted by an upper-case $F(\cdot)$, whereas the pmf is denoted by a lower-case $f(\cdot)$. The more-explicit notation, $F_X(\cdot)$ and $f_X(\cdot)$, becomes necessary when dealing with more than one random variable.

Comment. Both the pmf and cdf are defined for all $x \in \mathbb{R}$. For x not in the support of $f(x)$, $f(x) \equiv 0$.

Comment. The expectation operator, $\mathrm{E}[\cdot]$, can operate on any function of a random variable (discrete, continuous, or mixed) and is the constant defined by

$$\mathrm{E}[h(X)] \equiv -\int_{-\infty}^{0} \mathrm{P}(h(X) \le x)\mathrm{d}x + \int_{0}^{\infty} \mathrm{P}(h(X) > x)\mathrm{d}x,$$

for any function $h(\cdot)$.

Result. For any differentiable function $h(X)$ of random variable X (discrete, continuous, or mixed) that has $h(0) = 0$

$$\mathrm{E}[h(X)] \equiv -\int_{-\infty}^{0} \frac{\mathrm{d}h(a)}{\mathrm{d}a}\big|_{a=x} F(x)\mathrm{d}x + \int_{0}^{\infty} \frac{\mathrm{d}h(a)}{\mathrm{d}a}\big|_{a=x} (1 - F(x))\,\mathrm{d}x.$$

Comment. Alternatively for discrete random variables, the expectation operator can be computed

$$\mathrm{E}[h(X)] = \sum_{\text{all } x_i} h(x_i) f(x_i).$$

Definition. The variance is the *second moment about the mean*, or *moment of inertia*.

Definition. The variance for random variable $X(\cdot)$ (discrete, continuous, or mixed) is a constant defined by

$$\begin{aligned} \mathrm{V}[X] &\equiv \mathrm{E}[(X - \mu)^2] \\ &= \mathrm{E}[X^2] - (\mathrm{E}[X])^2. \end{aligned}$$

– $(\mathrm{E}[X])^2$ is often written as $\mathrm{E}^2[X]$.

- The variance is also commonly denoted by σ^2 or, more explicitly, by σ_X^2, or Var $[X]$.
- Var $[X]$ is a nonnegative constant whose units are the square of the units of $X(\cdot)$.

Comment. The variance for random variable $X(\cdot)$ (discrete, continuous, or mixed) is computed

$$\begin{aligned} V[X] = & - \int_{-\infty}^{0} P\left((X-\mu)^2 \le x\right) dx \\ & + \int_{0}^{\infty} P\left((X-\mu)^2 > x\right) dx \\ = & - \int_{-\infty}^{0} 2xF(x)dx + \int_{0}^{\infty} 2x\left(1-F(x)\right) dx \\ & - \left(-\int_{-\infty}^{0} F(x)dx + \int_{0}^{\infty} (1-F(x)) dx\right)^2. \end{aligned}$$

Definition. Alternatively for discrete random variables, the variance can be computed

$$\begin{aligned} V[X] & = \sum_{\text{all } x_i} (x_i - \mu_X)^2 f(x_i), \\ & = \sum_{i=1}^{n} x_i^2 f(x_i) - \mu^2. \end{aligned}$$

Definition. The *standard deviation* of $X(\cdot)$ is the constant $\sigma \equiv \sigma_X \equiv +\sqrt{V[X]}$.

Comment. The units of σ ($\equiv \sigma_X$) are the units of $X(\cdot)$.

Comment. For the remainder of this book the notation X, f, and F is used as a shorthand for $X(\cdot), f(\cdot)$, and $F(\cdot)$.

Definition. A random variable X has a *discrete uniform* distribution if each of the n numbers in its range, say $x_1, x_2, \ldots, x_n$, has equal probability.

Result. Suppose that X has a discrete uniform distribution on the $n > 1$ equally spaced numbers from a to b, and c is the distance between points, then

$$f(x) = 1/n \text{ for } x = a, a+c, a+2c, \ldots, b,$$

where $c = (b-a)/(n-1)$,

and $E[X] = \dfrac{a+b}{2}$ and $V[X] = \left(\dfrac{(b-a)^2}{12}\right)\left(\dfrac{n+1}{n-1}\right).$

Definition. An *indicator random variable*, X, takes on values of one or zero, depending on whether an event occurs. That is, $X(\omega) = 1$ if $\omega \in E$, and $X(\omega) = 0$ if $\omega \notin E$.

– Comment. $E[X] = P(E)$
– Comment. $\text{Var}[X] = P(E)(1 - P(E))$.

Definition. A *Bernoulli* trial is a random experiment that has exactly two outcomes (often called "success" and "failure").

Definition. A sequence of *Bernoulli trials* (or *Bernoulli process*) has three properties.

1. Each trial is a Bernoulli trial.
2. Each trial has $P(\{\text{success}\}) = p$, which is a constant.
3. Each trial is independent of every other trial.

Definition. A *Bernoulli random variable* is a random variable that assigns 0 to the outcome "failure," and 1 to the outcome "success" on the Bernoulli trial; i.e., $X(failure) = 0$, and $X(success) = 1.$

Definition. A *geometric random variable* is a random variable that counts the number of Bernoulli trials until the first success.

- $F_X(x) = 1 - (1-p)^x$, for $i = 1, 2, 3, \ldots$, where p is the probability of "success" on any particular Bernoulli trial. (The pmf is listed in Table I.)
- Comment. The geometric distribution is the only discrete distribution with the *memoryless property*; i.e.,
$$P(X > x + c \mid X > c) = P(X > x),$$
for all $x > 0$ and $c > 0$.

Definition. A *binomial experiment* is an experiment composed of n Bernoulli trials.

Definition. A *binomial random variable*, W, is the number of successes in a binomial experiment. Let X_i be the Bernoulli random variable for trial i; thus
$$W = \sum_{i=1}^{n} X_i.$$

Definition. An ordering of r elements from a set of n elements is called a *permutation*.

Definition. A selection (without regard to order) of r elements from a set of n elements is called a *combination*.

Result. A set of n elements has
$n! = n \times (n-1) \times (n-2) \times \cdots \times 2 \times 1$ permutations.
(The empty set has $0! = 1$ permutation.)

Result. The number of permutations of r elements from a set of n elements is
$$P_r^n \equiv \frac{n!}{(n-r)!} \text{ for } r = 0, 1, \ldots, n.$$

Result. The number of combinations of r elements from a set of n elements is

$$C_r^n \equiv \binom{n}{r} \equiv \frac{n!}{r!(n-r)!} = \frac{P_r^n}{r!} \text{ for } r = 0, 1, \ldots, n.$$

Definition. If X is the binomial random variable associated with a binomial experiment, then X has the *binomial distribution*

$$f_X(i) \equiv \mathrm{P}(X = i) = \binom{n}{i} p^i (1-p)^{n-i},$$

for $i = 0, \ldots, n$, and 0 elsewhere.

Comment. "$X \sim \mathrm{binom}(n, p)$" is read "$X$ has a binomial distribution with parameters n and p."

Definition (not precise). Suppose that a random variable that counts certain events has unit increments (simultaneously occurring events cannot happen) occurring at random throughout a real-number interval typically of time (or space) of length t. Then the expected event-count increases at a rate that is proportional to t. The proportionality constant is called the process rate and is commonly denoted by λ.

The random experiment is called a (*homogeneous*) *Poisson process* if and only if the time interval can be partitioned into equal-length non-overlapping subintervals of small enough length that the following three conditions hold.

1. The probability of more than one event-count increment in a time subinterval is approximately zero.

2. The probability of one event-count increment in a time subinterval is the same for all equal-length time subintervals and approximately proportional to the length of the time subintervals with proportionality constant λ.

3. The event-count increments in each time subinterval are independent of other non-overlapping time subintervals.

These conditions are strictly true in the limiting sense; i.e., as the size of the time subintervals goes to zero.

(Example: Customer arrivals. Condition (1) says that customers arrive separately; Condition (2) says that the arrival rate, λ, is constant across time (that is, homogeneous) and that the probability of a customer arriving during a two-second subinterval is approximately twice the probability of a customer arriving during a one-second subinterval; Condition (3) says that the number of customers in one time subinterval is not useful in predicting the number of customers in another time subinterval.)

Comment. The Poisson process is a *memoryless* process; i.e., the number of event-count increments in future intervals is independent of the number of event-count increments occurring in the past.

Comment. "$N \sim \text{Poisson}(\mu)$" is read "$N$ is a Poisson random variable having mean μ."

Comment. "$N(t) \sim \text{Poisson}(\lambda t)$" is read "$N(t)$ is a Poisson process having rate λ."

Comment. For a Poisson process, $N(t)$, the associated number of count increments in the interval $(0, t)$, is a random variable that is Poisson with mean $\mu = \lambda t$.

Result: *Poisson Approximation to the Binomial Distribution.* The Poisson process is (asymptotically) equivalent to many, n, Bernoulli trials, each with a small probability of success, p. Therefore, the Poisson distribution with mean np is a good approximation to the binomial distribution when n is large and p is small; i.e.,
$$\text{binom}(n, p) \approx \text{Poisson}(np).$$

Definition: Hypergeometric Distribution. Consider a population of N items, of which K are successes (and $N - K$ are failures). The experiment is to choose, equally likely and without replacement, n items from the population. The random variable, say X, the number of the n items that are successes.

> Comment: Despite the usual convention that capital letters denote random variables, here N and K are constants.
>
> Result: The range of X is the set of integers with a minimum of $\max\{0, n + K - N\}$ and a maximum of $\min\{n, K\}$.
>
> Result: The hypergeometric pmf, mean, and variance are stated in Table 1.
>
> Comment: The roles of n and K are interchangeable. That is, the probability is unchanged if the experiment is to choose K items from a population of N items that contains n successes.

Comment. The binomial distribution with $p = K/N$ is a good approximation to the hypergeometric distribution when n is small compared to N.

Result. Let X is a discrete random variable with support $\mathcal{S}_X$ and let $Y = g(X)$, where $g(x)$ is a one-to-one function that maps $\mathcal{S}_X$ onto $\mathcal{S}_Y$. Let $g^{-1}(y)$ be the inverse function of $g(x)$, then the pmf for Y, a discrete random variable with support $\mathcal{S}_Y$, is

$$f_Y(y) \equiv f_X\left(g^{-1}(y)\right)$$

for $y \in \mathcal{S}_Y$.

Comment. If Y is a n-to-one transformation of X, then partition the support of X, $\mathcal{S}_X$, into n partitions, $\mathcal{S}_X^i$, $i = 1, 2, \ldots, n$, as needed to ensure that $Y = g(X)$ is a one-to-one function that maps each partition $\mathcal{S}_X^i$ onto a corresponding partition of the support of Y, $\mathcal{S}_Y^i$. The resulting pdf for Y is

$$f_Y(y) \equiv \sum_{i=1}^{n} f_X\left(g_i^{-1}(y)\right)$$

for $y \in \mathcal{S}_Y$.

Table I – Discrete Distributions

random variable	distribution name	range
X	general	$x_1, x_2, \ldots, x_n$
X	discrete uniform	$x_1, x_2, \ldots, x_n$
X	equally spaced uniform	$x = a, a+c, \ldots, b$
"# successes in one Bernoulli trial"	indicator or Bernoulli random variable	$x = 0, 1$
"# successes in n Bernoulli trials"	binomial (sampling with replacement)	$x = 0, 1, \ldots, n$
"# successes in a sample of size n from a population of size N containing K successes"	hyper-geometric (sampling without replacement)	$x = (n - (N-K))^+, \ldots, \min\{K, n\}$ and integer

probability mass function	mean (constant)	variance (constant)
$\equiv P(\{\omega\|X(\omega)=x\})$ $\equiv f(x)$ $\equiv f_X(x)$ $P(X=x)$	$\sum_{i=1}^{n} x_i f(x_i)$ $\equiv \mu \equiv \mu_X$ $\equiv E[X]$	$\sum_{i=1}^{n} (x_i - \mu)^2 f(x_i)$ $\equiv \sigma^2 = \sigma_X^2$ $\equiv V[X] \equiv \text{Var}[X]$ $= E[X^2] - \mu^2$
$1/n$	$\sum_{i=1}^{n} x_i/n$	$(\sum_{i=1}^{n} x_i^2/n) - \mu^2$
$1/n$ where $n = (b-a+c)/c$	$(a+b)/2$	$c^2(n^2-1)/12$
$p^x(1-p)^{1-x}$	p	$p(1-p)$
$C_x^n p^x (1-p)^{n-x}$	where np	$p \equiv P(\text{``success''})$ $np(1-p)$
$C_x^K C_{n-x}^{N-K} / C_n^N$	where np where	$p \equiv P(\text{``success''})$ $np(1-p)\frac{(N-n)}{(N-1)}$ $p = K/N$

Table I – Discrete Distributions - Cont'd

random variable	distribution name	
"# Bernoulli trials until first success"	geometric	$x = 1, 2, \ldots$
"# failures prior to first success in a Bernoulli process"	shifted or modified geometric	$x = 0, 1, 2, \ldots$
"# Bernoulli trials until r^{th} success"	negative binomial	$x = r, r+1, \ldots$
"# of counts in time t from a Poisson process having rate λ"	Poisson	$x = 0, 1, \ldots$

probability mass function	mean (constant)	variance (constant)		
$p(1-p)^{x-1}$	$1/p$	$(1-p)/p^2$		
	where	$p \equiv \text{P(``success'')}$		
$p(1-p)^x$	$(1-p)/p$	$(1-p)/p^2$		
	where	$p \equiv \text{P(``success'')}$		
$C_{r-1}^{x-1} p^r (1-p)^{x-r}$	r/p	$r(1-p)/p^2$		
	where	$p \equiv \text{P(``success'')}$		
$e^{-\mu} \mu^x / x!$	μ	μ		
	where	$\mu = \lambda t$		

Continuous Random Variables and Probability Distributions

Comment. Concepts of, and notations for, the case of continuous random variables are analogous to the discrete random variable case, with integrals replacing sums.

Technical Issue. Since uncountable sample spaces contain an uncountable number of outcomes then it follows that some subsets of the sample space consisting of a single member must have probabilities that map to 0. If the subsets of the sample space consisting of single members each had a non-negative probability, then the union of all such singleton sets (the sample space, Ω) could not map to 1, and that would violate the axiom of probability that specifies that $P(\Omega) = 1$.[2]

Thus for uncountable sample spaces (and therefore continuous random variables) we define non-zero probabilities only for subsets of the sample space (events) that correspond to the continuous random variable taking some value that is in an interval, or the union of a set of intervals, on the real number line.

Another way to think about this is that we consider the domain of the probability set function associated with a continuous random variable as the family of subsets of the sample space that are implicitly defined by the random variable having some value in an interval on the real number line that is open on the left and closed on the right, or a union of such intervals.

---

[2] Although the union over an uncountable set of subsets can be formed, a sum is not defined for an uncountable index set; thus Axiom 3 fails. The structure of a "sigma field" is required to formally define the family of events for which probability can be defined. See any mathematical probability and statistics text for the technical details.

So we have

- For uncountable Ω, $P(\{\omega\}) \neq 0$, only for ω being in a countable subset of Ω.
- For a continuous random variable, X, $P(X = x) = 0$, for all $x \in \mathbb{R}$.
- For a continuous random variable, X, $P(X \in A)$ is possibly positive only if A is a subset of $\mathbb{R}$ that can be constructed from a finite number of unions and intersections of intervals (that are open on the left and closed on the right) of the real-number line; i.e., $A \equiv \{x \in (a, b]\}$, for real numbers $a \leq b$.

Definition. For a continuous random variable X, the *probability density function*, f (or f_X), often abbreviated *pdf*, is a function satisfying

1. $f(x) \geq 0$ for every real number x,
2. $\int_{-\infty}^{\infty} f(x)\, dx = 1$, and
3. $P(a < X \leq b) = \int_a^b f(x)\, dx$ for $a \leq b$, and $a, b \in \mathbb{R}$.

Result. For a continuous random variable X, the cdf is $F(x) \equiv F_X(x) \equiv P(X \leq x) = \int_{-\infty}^{x} f(u)\, du$, and the pdf is $f(x) \equiv f_X(x) \equiv \frac{d P(X \leq u)}{du}\big|_{u=x} = d(\int_{-\infty}^{x} f(u)\, du)/dx$. (This result is via the Fundamental Theorem of Calculus).

Comment. For all real numbers, x,
$$F_X(x) \equiv P(\{\omega | X(\omega) \leq x\})$$
and
$$f_X(x) \equiv \frac{d P(\{\omega | X(\omega) \leq u\})}{du}\Big|_{u=x}.$$

Observe. The pdf is analogous to the pmf.

- If X is a discrete random variable, then
$$P(a < X \leq b) \equiv \sum_{x \in (a,b]} f_X(x),$$
and
$$P(a < X \leq b) \equiv \int_a^b f_X(x)\,dx.$$

- The pmf is a probability; the pdf is not a probability.
- The pdf is a derivative of a probability; therefore, the pdf must be integrated to obtain a probability.

Result. If X is continuous, then $P(X = x) = 0$ for every real number x.

Comment. This is consistent with the elementary calculus result that $\int_x^x f(u)\,du \equiv 0$, for every real number x. Therefore, $P(x < X) = P(x \leq X)$, which is different from the discrete case.

Definition. The *mean* (or *first moment* or *first raw moment*, or *first power moment*, or the *expectation*) of a continuous random variable X is the constant
$$E[X] \equiv -\int_{-\infty}^{0} P(X \leq x)\,dx + \int_{0}^{\infty} P(X > x)\,dx$$
$$= -\int_{-\infty}^{0} F(x)\,dx + \int_{0}^{\infty} (1 - F(x))\,dx.$$

When the pdf of X exists another form of the mean is
$$E[X] = \int_{-\infty}^{\infty} x f(x)\,dx.$$

Comment. Alternative notation for $\mathrm{E}[X]$ is μ, or the more explicit μ_X.

Definition. For positive values of n, the nth raw moment (or n^{th} power moment) of a continuous random variable X is the constant

$$\mathrm{E}[X^n] \equiv -\int_{-\infty}^{0} nx^{n-1}\mathrm{P}(X \leq x)\,\mathrm{d}x + \int_{0}^{\infty} nx^{n-1}\mathrm{P}(X > x)\,\mathrm{d}x$$

$$= -\int_{-\infty}^{0} nx^{n-1}F(x)\,\mathrm{d}x + \int_{0}^{\infty} nx^{n-1}(1 - F(x))\,\mathrm{d}x.$$

When the pdf of X exists another form of the nth raw moment is

$$\mathrm{E}[X^n] = \int_{-\infty}^{\infty} x^n f(x)\,\mathrm{d}x.$$

Definition. For any function $g(\cdot)$ of a continuous random variable X, the constant

$$\mathrm{E}[g(X)] \equiv -\int_{-\infty}^{0} \mathrm{P}(g(X) \leq x)\,\mathrm{d}x$$

$$+ \int_{0}^{\infty} \mathrm{P}(g(X) > x)\,\mathrm{d}x.$$

Result. For any differentiable function $g(\cdot)$ of a continuous random variable X that has $g(0) = 0$ we have constant

$$\mathrm{E}[g(X)] \equiv -\int_{-\infty}^{0} \mathrm{P}(g(X) \leq x)\,\mathrm{d}x$$

$$+ \int_0^\infty P(g(X) > x)\,dx.$$

$$= -\int_{-\infty}^0 \frac{d\,g(a)}{da}\Big|_{a=x} (F(x))\,dx$$

$$+ \int_0^\infty \frac{d\,g(a)}{da}\Big|_{a=x} (1 - F(x))\,dx.$$

Comment. When the pdf of X exists, $E[g(X)]$ can be written

$$E[g(X)] = \int_{-\infty}^\infty g(x)f(x)\,dx.$$

Comment. For most of the remainder of this book expectations are shown using the probability-density function; i.e., we will assume that a density function exits.

Definition. The *variance* of a random variable X is the constant $V[X] \equiv E[(X - \mu)^2]$.
(Unchanged from the discrete case, and a special case of the expectation of a differentiable function listed above.)

Result. For a continuous random variable X

$$\sigma^2 \equiv \sigma_X^2 \equiv V[X]$$

$$\equiv \int_{-\infty}^\infty (x - \mu)^2 f(x)\,dx$$

$$= \int_{-\infty}^\infty x^2 f(x)\,dx - \mu^2$$

$$\begin{aligned} &= \mathrm{E}[X^2] - \mu^2 \\ &= \mathrm{E}[X^2] - \mathrm{E}^2[X]. \end{aligned}$$
(Recall that $(\mathrm{E}[X])^m$ is also written as $\mathrm{E}^m[X]$.)

Comment: Expectation is a linear operator; thus $\mathrm{E}[g(X)] \neq g(\mathrm{E}[X])$ for nonlinear $g(\cdot)$ and $\mathrm{E}[g(X)] = g(\mathrm{E}[X])$ for linear $g(\cdot)$. For example,

- $\mathrm{E}[a + bX] = a + b\mathrm{E}[X]$, and
- $\mathrm{E}[X^2] \neq (\mathrm{E}[X])^2$
  (if this were true then all variances would equal 0).

Definition. The *standard deviation* of the random variable X is the constant $\sigma \ (\equiv \sigma_X) \equiv +\sqrt{\mathrm{V}[X]}$.
(Unchanged from the discrete case.)

Definition. A random variable X has a *continuous uniform* (or *rectangular*) distribution if its pdf forms a rectangle with base $[a, b]$.

Result. If the random variable X has a continuous uniform distribution on $[a, b]$, then

- $\mathrm{E}[X] = (a + b)/2$, and
- $\mathrm{V}[X] = (b - a)^2/12$.

(Compare to the discrete uniform distribution.)

Definition. A random variable X has a triangular distribution if its pdf forms a triangle with base $[a, b]$ and mode at m, where $a \leq m \leq b$.

Result. If X has a triangular distribution on $[a, b]$, with mode at m, then

- $\mathrm{E}[X] = (a + m + b)/3$, and
- $\mathrm{V}[X] = ((b - a)^2 - (m - a)(b - m))/18$.

Definition. The *normal* (or *Gaussian*) distribution (see Table II) with mean μ (a constant) and standard deviation σ (a constant)

- is often an adequate model for the sum of many random variables
  (as a result of the "Central Limit Theorem"),
- has a symmetric bell-shaped pdf centered at the constant μ, points of inflection at the constants $\mu \pm \sigma$, and range $\mathbb{R}$,
- is the only famous distribution for which the commonly used notation μ and σ are used directly as the distribution's mean and standard deviation.
- "$X \sim N(\mu, \sigma^2)$" is read "X is normally distributed with mean μ and variance σ^2."

Definition. The *standard normal* distribution is the special case of the normal distribution where $\mu = 0$ and $\sigma = 1$. Random variables having this distribution are often denoted by Z, its cdf by $\Phi(\cdot)$, and its pdf by $\phi(\cdot)$.

Result. To convert between general and standardized normal distributions, use

- If $X \sim N(\mu, \sigma^2)$, then $(X - \mu)/\sigma \sim N(0, 1)$.
- If $Z \sim N(0, 1)$, then $(\mu + \sigma Z) \sim N(\mu, \sigma^2)$.

Probability calculations for the normal distribution are not closed form. Use numerical methods, approximations, or tabled values.

Comment. Most elementary probability and statistics books provide these tables. In this book we provide a Microsoft EXCEL code and a convenient one-line approximation.

- For $X \sim N(\mu, \sigma^2)$, relate p and x_p, where $p = P(X \leq x_p) \equiv F_X(x_p)$.
    - Microsoft EXCEL:
        - Given p, use NORMINV to compute x_p.
        - Given x_p use NORMDIST to compute p.
    - Normal probabilities can be converted to standard normal probabilities using
    $$F_X(x_p) \equiv P(X \leq x_p)$$
    $$= P\left(\underbrace{\frac{X-\mu}{\sigma}}_{\text{random variable}} \leq \underbrace{\frac{x_p-\mu}{\sigma}}_{\text{constant}}\right)$$
    $$= P(Z \leq z_p) \equiv \Phi(z_p),$$
    where $z_p = (x_p - \mu)/\sigma$ is the z-value obtained by *standardizing* $X = x_p$.
- For $Z \sim N(0, 1)$, relate p and z_p, where $p = P(Z \leq z_p) \equiv \Phi(z_p)$.
    - Microsoft Excel:
        - Given p, use NORMSINV to compute z_p.
        - Given z_p, use NORMSDIST to compute p.
- Schmeiser One-Line Approximations:
    - $z_p \approx (p^{0.135} - (1-p)^{0.135})/0.1975$
    - $p \approx (1 + \exp(-z_p(1.5966 + (z_p^2/14))))^{-1}$
- A Sketch of the normal density visually reveals the approximate relevant area. As an aid, remember
    - $P(\mu - \sigma < X < \mu + \sigma) \approx 0.68$
    - $P(\mu - 2\sigma < X < \mu + 2\sigma) \approx 0.95$
    - $P(\mu - 3\sigma < X < \mu + 3\sigma) \approx 0.997$.

Definition. The *order statistics* from a random sample, $X_1, X_2, \ldots, X_n$, are the random variables $X_{(1)} \leq X_{(2)} \leq \ldots \leq X_{(n)}$. That is $X_{(1)}$ is the random variable that is the minimum from the random sample, and $X_{(n)}$ is the random variable that is the maximum from the random.

Caution. Just as the numerical value x_i is a real number that is the value of the ith observation of the set of observations $x_1, x_2, \ldots, x_n$, so too the numerical value $x_{(i)}$ is the value of the ith-smallest observation from the set of observations $x_1, x_2, \ldots, x_n$.

A graphical test for normality. Given a random sample of size n, plot each of the order-statistic observations, $x_{(j)}$, against its *standardized normal score* z_p, the $P_j = (j - 0.5)/n$ quantile (real numbers). (Some texts prefer $P_j = j/(n+1)$). If the resulting curve is approximately a straight line, then the sample data are consistent with the observations having been drawn from a normal distribution.

Result. *Normal approximation to the binomial distribution.* The normal $N(np, np(1-p))$ distribution is a good approximation to the binomial(n,p) distribution when $\min[\mu_X, n - \mu_X] = n \min[p, (1-p)]$ is "large." (A common value for "large" is 5; the approximation is asymptotically exact. See also "continuity correction.")

Result. *Normal approximation to the Poisson distribution.* The $N(\mu_X, \mu_X)$ distribution is a good approximation to the Poisson(μ_X) distribution when μ_X is "large." (A common value for "large" is 5; the approximation is asymptotically exact. See also "continuity correction.")

Definition. *Continuity correction.* Rewriting an event to reduce the error from approximating a discrete distribution with a continuous distribution.

Let X be a discrete random variable. For simplicity, assume that the possible values of X are integers. Then, for integer constants a and b,

$$P(a \leq X \leq b) = P(a - 0.5 \leq X \leq b + 0.5).$$

The continuity-correction approximation of $P(a \leq X \leq b)$ is the continuous-distribution's value of $P(a - 0.5 \leq X \leq b + 0.5)$. (The correction is crucial when $a = b$, but less important as $b - a$ increases.)

Result. From any time t, let X denote the time until the next count of a Poisson Process having rate λ. Then X has an exponential distribution with mean $1/\lambda$. (Here t can be any time, including zero or the time of a particular count.) (See Table II.)

Result. For any time t, let X denote the time until the rth count of a Poisson Process having rate λ. Then X is the sum of r independent exponential random variables and X has an Erlang distribution with parameters r and λ. (See Table II.)

Analogy. The exponential distribution is analogous to the geometric distribution; the Erlang distribution is analogous to the negative-binomial distribution.

Definition. The *gamma* distribution is the generalization of the Erlang distribution in which the parameter $r > 0$ is not necessarily an integer. (See Table II.)

Result. If Y is an exponential random variable with mean 1, then $X = \delta Y^{1/\beta}$ is *Weibull* with parameters $\delta > 0$ and $\beta > 0$. (See Table II.)

Definitions. Three types of distribution parameters:

- A *location* parameter a is additive constant: $Y = a + X$. The distribution of Y is identical to that of X except that its location (and thus its mean) is shifted a units to the right. The variances for X and Y are identical.
- A *scale* parameter b is multiplicative constant: $Y = bX$. The distribution of the random variable Y is identical to that of the random variable X, except that each unit of X is b units of Y. This change in units is reflected in all moments.
- A *shape* parameter c is a constant in a nonlinear transformation of X: $Y = g(X; c)$, where the function $g(\cdot)$ is nonlinear in c. The distributions of Y and X (can) have different shapes, means, and variances.

Result. *Chebyshev's Inequality.* The probability that a random variable X differs from its mean by at least c standard deviations is no more than $1/c^2$. Notationally,

$$P(|X - \mu| \geq c\sigma) \leq 1/c^2$$

for every constant $c > 0$. Equivalently, the inequality can be written as

$$P(\mu - c\sigma < X \leq \mu + c\sigma) \geq 1 - 1/c^2.$$

Comment. For example, every distribution has at least $8/9 = 88.9\%$ of the probability within three standard deviations of the mean. Chebyshev's inequality holds for all distributions, but seldom is a good approximation for a particular distribution. Chebyshev's inequality provides tighter bounds for distributions having symmetric pdfs than for distributions having asymmetric pdfs. For comparison, recall that about 99.7% of the probability is within three standard deviations of the mean for the normal distribution.

The inequality is trivially true for $0 < c < 1$. This inequality is just one from a family of Chebyshev inequalities that are found in more advanced probability texts.

Chebyshev is spelled various ways by different authors, including Tchebychev, Tchebyshev, Tchebysheff... .

Result. Let X is a continuous random variable with support $\mathcal{S}_X$ and let $Y = g(X)$, where $g(x)$ is a one-to-one differentiable function on $\mathcal{S}_X$. Let $g^{-1}(y)$ be the inverse function of $g(x)$, then the pdf for Y, a continuous random variable with support $\mathcal{S}_Y$, is

$$f_Y(y) \equiv f_X\left(g^{-1}(y)\right)\left|\frac{dg^{-1}(y)}{dy}\right|$$

for $y \in \mathcal{S}_Y$.

Comment. If Y is a n-to-one transformation of X, then partition the support of X, $\mathcal{S}_X$, into n partitions, $\mathcal{S}_X^i$, $i = 1, 2, \ldots, n$, as needed to ensure that $Y = g(X)$ is a one-to-one differentiable function that maps each partition $\mathcal{S}_X^i$ into the corresponding partition of the support of Y, $\mathcal{S}_Y^i$. The resulting pdf for Y is

$$f_Y(y) \equiv \sum_{i=1}^{n} f_X\left(g_i^{-1}(y)\right)\left|\frac{dg_i^{-1}(y)}{dy}\right|$$

for $y \in \mathcal{S}_Y$.

Table II – Continuous Distributions

rand. var.	dist. name	range	probability distrib. func.
X	general	$(-\infty, \infty)$	$P(X \leq x)$ $\equiv F(x)$ $\equiv F_X(x)$
X	continuous uniform	$[a, b]$	$(x-a)/(b-a)$
X	triangular	$[a, b]$	$(x-a)f(x)/2$ if $x \leq m$, else $1 - (b-x)f(x)/2$
sum of random variables	normal or Gaussian	$(-\infty, \infty)$	via tables*, or EXCEL, or Schmeiser approx.
time to Poisson count 1	exponential	$[0, \infty)$	$1 - e^{-\lambda x}$

* See any elementary probability and statistics textbook for tabled values of the standard normal cdf.

probability density func.	mean (constant)	variance (constant)
$\frac{dF(y)}{dy}\big|_{y=x}$ $\equiv f(x)$ $\equiv f_X(x)$	$\int\limits_{-\infty}^{\infty} x f(x)\,\mathrm{d}x$ $\equiv \mu \equiv \mu_X$ $\equiv \mathrm{E}[X]$	$\int\limits_{-\infty}^{\infty} (x-\mu)^2 f(x)\,\mathrm{d}x$ $\equiv \sigma^2 \equiv \sigma_X^2$ $\equiv \mathrm{V}[X]$ $= \mathrm{E}[X^2] - \mu^2$
$1/(b-a)$	$(a+b)/2$	$(b-a)^2/12$
$\frac{2(x-d)}{(b-a)(m-d)}$ ($d=a$ if $x \leq m$, else $d=b$)	$\frac{(a+m+b)}{3}$	$\frac{(b-a)^2-(m-a)(b-m)}{18}$
$\frac{\exp\left(-\frac{1}{2}((x-\mu/\sigma)^2\right)}{(\sqrt{2\pi}\sigma)}$	μ	σ^2
$\lambda e^{-\lambda x}$	$1/\lambda$	$1/\lambda^2$

Table II – Continuous Distributions – Cont'd.

rand. var.	dist. name	range	probability distrib. func.
time to Poisson count r	Erlang	$[0, \infty)$	$\sum_{k=r}^{\infty} (\lambda x)^k e^{-\lambda x}/k!$
lifetime	gamma	$[0, \infty)$	numerical
lifetime	Weibull	$[0, \infty)$	$1 - e^{-(x/\delta)^\beta}$
lifetime	lognormal	$[0, \infty)$	numerical

Definition. For any $r > 0$, the *gamma function* is
$$\Gamma(r) \equiv \int_0^\infty x^{r-1} e^{-x}\, dx.$$

Result. $\Gamma(r) = (r-1)\,\Gamma(r-1)$. In particular, if r is a positive integer, then $\Gamma(r) = (r-1)!$.

Comment. The exponential distribution is the only continuous memoryless distribution. That is,

$$P(X > x + c \,|\, X > x) = P(X > c)$$

for all $x > 0$ and $c > 0$.

Definition. A *lifetime* distribution is continuous with range $[0, \infty)$.

Modeling lifetimes. Some useful lifetime distributions are the exponential, Erlang, gamma, lognormal, and Weibull.

probability density func.	mean (constant)	variance (constant)
$\frac{\lambda^r x^{r-1} e^{-\lambda x}}{(r-1)!}$	r/λ	r/λ^2
$\frac{\left(\lambda^r x^{r-1} e^{-\lambda x}\right)}{\Gamma(r)}$	r/λ	r/λ^2
$\frac{(\beta/\delta)(x/\delta)^{\beta-1}}{\exp\left((x/\delta)^\beta\right)}$	$\delta\,\Gamma(1+\frac{1}{\beta})$	$\delta^2\,\Gamma(1+\frac{2}{\beta}) - \mu^2$
$\frac{\exp\left(-\ln(y-\mu)\right)^2/(2\sigma^2)}{\left(y\sigma\sqrt{2\pi}\right)}$	$e^{\mu+\left(\sigma^2/2\right)}$	$e^{2\mu+\sigma^2}\left(e^{\sigma^2}-1\right)$

Random Vectors and Joint Probability Distributions

Comment. The topic is now two (or more) random variables defined on the same sample space.

Notation. Lower-case letters denote constants and upper-case letters denote random variables, except that lower-case f denotes pmf's and pdf's, and upper-case F denotes cdf's. Subscripts in joint and conditional-function names become important.

Definition. A *random vector*, say $(X_1, X_2, \ldots, X_n)$, is a vector whose components are (scalar) random variables defined on a common sample space.

Comment. Each random variable can be discrete, continuous, or mixed. Discrete random variables require summation and continuous random variables require integration.

Definition. The joint probability distribution of a random vector $(X_1, X_2, \ldots, X_n)$ is a description (in whatever form) of the likelihoods associated with the possible values of $(X_1, X_2, \ldots, X_n)$.

Comment. When $n = 2$, the two random variables are often denoted by X and Y. Definitions are given here only for $n = 2$; they extend to general values of n by analogy.

Definition. The *joint cumulative distribution function (cdf)* of the random vector (X, Y) is, for all real numbers x and y,
$$F_{X,Y}(x, y) \equiv \mathrm{P}(X \leq x, Y \leq y),$$
where the comma denotes intersection and is read as "and."

Result. For all real numbers $a \leq b$ and $c \leq d$,

$$P(a < X \leq b, c < Y \leq d)$$
$$= F_{X,Y}(b,d) - F_{X,Y}(b,c) - F_{X,Y}(a,d) + F_{X,Y}(a,c).$$

Definition. The *joint probability mass function (pmf)* of the discrete random vector (X, Y) is

$$f_{X,Y}(x, y) = P(X = x, Y = y),$$

for all real numbers x and y.

Result. Every joint pmf satisfies

1. $f_{X,Y}(x, y) \geq 0$, for all real numbers x and y, and
2. $\sum\limits_{\text{all } x} \sum\limits_{\text{all } y} f_{X,Y}(x, y) = 1.$

Definition. The *marginal distribution* of X is the distribution of X alone, unconditional on Y.

Comment. Independence of random variables is analogous to independence of events.

Result. The marginal cdf of X is, for every real number x,

$$F_X(x) = \lim_{y \to \infty} F_{X,Y}(x, y).$$

Likewise

$$F_Y(y) = \lim_{x \to \infty} F_{X,Y}(x, y).$$

Result. If (X, Y) is a discrete random vector, then for every $x \in \mathbb{R}$ the marginal pmf of X is

$$f_X(x) = P(X = x) = \sum_{\text{all } y} f_{X,Y}(x, y).$$

Definition. Discrete random vector (X,Y) with joint pmf $f_{X,Y}(x,y)$ has support where $f_{X,Y}(x,y) > 0$.

Result. If the discrete random vector (X,Y) has pmf $f_{X,Y}(x,y)$ and has support (x_i, y_j), where $i \in \mathbf{I}$, and $j \in \mathbf{J}$, then for $x_{i-1} \leq x_i$ and $y_{j-1} \leq y_j$, $f_{X,Y}(x_i, y_j)$
$$\begin{aligned}
&= \mathrm{P}(X = x_i, Y = y_j) \\
&= \mathrm{P}(X = x_i, Y \leq y_j) - \mathrm{P}(X = x_i, Y \leq y_{j-1}) \\
&= \mathrm{P}(X \leq x_i, Y \leq y_j) - \mathrm{P}(X \leq x_{i-1}, Y \leq y_j) \\
&\quad - \mathrm{P}(X \leq x_i, Y \leq y_{j-1}) + \mathrm{P}(X \leq x_{i-1}, Y \leq y_{j-1}) \\
&= F(x_i, y_j) - F(x_{i-1}, y_j) \\
&\quad - F(x_i, y_{j-1}) + F(x_{i-1}, y_{j-1}).
\end{aligned}$$

Result. The marginal mean of X is the constant
$$\mathrm{E}[X] \equiv \mu_X \equiv \sum_{\text{all } x} x f_X(x) = \sum_{\text{all } x} \sum_{\text{all } y} x f_{X,Y}(x,y).$$

Result. The marginal variance of X is the constant
$$\begin{aligned}
\mathrm{V}[X] &\equiv \sigma_X^2 \equiv \sum_{\text{all } x} (x - \mu_X)^2 f_X(x) \\
&= \sum_{\text{all } x} \sum_{\text{all } y} (x - \mu_X)^2 f_{X,Y}(x,y).
\end{aligned}$$

Comment. The marginal standard deviation of the random variable X is $\sigma_X = +\sqrt{\sigma_X^2}$.

Comment. The means, variances, and standard deviations are constants. The units of the mean and standard deviation are the units of the random variable X. The variance of the random variable X has units of X^2.

Comment. Analogous results hold for the marginal cdf, pmf, mean and variance and standard deviation of the random variable Y.

Definition. If (X, Y) is a discrete random vector with joint pmf $f_{X,Y}$, then the *conditional pmf* of Y given $X = x$ is

$$f_{Y|X}(y|x) \equiv \frac{f_{X,Y}(x,y)}{f_X(x)},$$

if $f_X(x) > 0$, and undefined otherwise.

Comment. Compare to $P(A|B) = \frac{P(A \cap B)}{P(B)}$.

Comment. Alternate ways of writing the conditional pmf, cdf, and expectations are common. One common alternative for the conditional pmf is

$$f_{Y|X}(y|x) \equiv f_{Y|X=x}(y)$$

The notation "$f_{Y|X}(y|x)$" is probably the most explicit and is the notation that used for the remainder of this book. The notation "$f(y|x)$" is commonly used in probability texts when there is no ambiguity in meaning in the context of the model.

Comment. The conditional pmf of X given $Y = y$ is defined analogously.

Result. For every real number x at which $f_{Y|X}(y|x)$ is defined

1. $f_{Y|X}(y|x) \geq 0$ for every real number y,
2. $\sum\limits_{\text{all } y} f_{Y|X}(y|x) = 1$, and
3. $f_{Y|X}(y|x) = P(Y = y | X = x)$, for every real number y and every real number x.

−54−

Result.
$$E[Y|X=x] \equiv \mu_{Y|X=x} = \sum_{\text{all } y} y\, f_{Y|X}(y|x)$$

and

$$\begin{aligned}
V[Y|X=x] &\equiv \sigma^2_{Y|X=x} = \sum_{\text{all } y} (y - \mu_{Y|X=x})^2 f_{Y|X}(y|x) \\
&= E[Y^2|X=x] - \mu^2_{Y|X=x},
\end{aligned}$$

where $E[Y^2|X=x] = \sum_{\text{all } y} y^2 f_{Y|X}(y|x)$.

Comment. The conditional mean and conditional variance of Y given $X = x$ are constants if a specific real number is given for x. The conditional mean and conditional variance of Y given $X = x$ are functions of the dummy variable x if no specific real number is given for x.

Caution. The dummy variable x is not a random variable.

Result. If (X, Y) is a discrete random vector, then

$$f_{X,Y}(x,y) = f_{X|Y}(x|y) f_Y(y) = f_{Y|X}(y|x) f_X(x)$$

for all x and y.

Definition. If (X, Y) is a random vector, then X and Y are *independent* if and only if for every x and y

$$f_{X,Y}(x,y) = f_X(x) f_Y(y).$$

Result. If (X, Y) is a random vector, then the following four statements are equivalent; that is, either none are true or all are true.

1. X and Y are independent.
2. $f_{Y|X}(y|x) = f_Y(y)$ for all x and y with $f_X(x) > 0$.
3. $f_{X|Y}(x|y) = f_X(x)$ for all x and y with $f_Y(y) > 0$.
4. $P(X \in A, Y \in B) = P(X \in A)P(Y \in B)$ for all subsets A and B of $\mathbb{R}$.

Caution.

- $P(E_1|E_2) = P(E_1)$ where $P(E_2) > 0$, means that events E_1 and E_2 are independent.
- $f_{Y|X}(y|x) = f_Y(y)$ where $f_X(x) > 0$ for all x and y, means that X and Y are independent.
- $F_{Y|X}(y|x) = F_Y(y)$ where $f_X(x) > 0$ for all x and y, means that X and Y are independent.

And

- $P(E_1|E_2) \equiv P(E_1 \cap E_2)/P(E_2)$, where $P(E_2) > 0$.

- $f_{Y|X}(y|x) \equiv f_{XY}(x,y)/f_X(x)$, where $f_X(x) > 0$ for all x and y.

However: $F_{Y|X}(y|x) \neq F_{X,Y}(x,y)/F_X(x)$, where $F_X(x) > 0$, for all x and y.

Definition. A *multinomial experiment* is composed of n trials satisfying

1. each trial has exactly one of k outcomes;
2. the probability of the ith outcome on any trial is p_i for $i = 1, 2, \ldots, k$, and
3. the trials are independent.

–56–

Definition. In a multinomial experiment, let X_i denote the number of trials that result in the ith outcome for $i = 1, 2, \ldots, k$. (Thus $X_1 + X_2 + \cdots + X_k = n$.) The random vector $(X_1, X_2, \ldots, X_k)$ has a *multinomial distribution* with joint pmf

$$P(X_1 = x_1, X_2 = x_2, \ldots, X_n = x_n)$$
$$= \frac{n!}{x_1! x_2! \cdots x_k!} p_1^{x_1} p_2^{x_2} \cdots p_k^{x_k},$$

when each x_i is a nonnegative integer and $x_1 + x_2 + \cdots + x_k = n$; and zero elsewhere.

Result. If the random vector $(X_1, X_2, \ldots, X_k)$ has a multinomial distribution, the marginal distribution of X_i is binomial with parameters n and p_i for $i = 1, 2, \ldots, k$.
(And therefore $E[X_i] = np_i$ and $V[X_i] = np_i(1 - p_i)$.)

Comment. The topic is now two (or more) continuous random variables defined on the same sample space. All previous cdf results hold for both discrete and continuous random variables; they are not repeated here.

Definition. The *joint probability density function (pdf)* of the continuous random vector (X, Y) is, for real numbers x and y, denoted by $f_{X,Y}(x, y)$ and satisfies

1. $f_{X,Y}(x, y) \geq 0$ for all real numbers x and y, and
2. $\int_{-\infty}^{\infty} \int_{-\infty}^{\infty} f_{X,Y}(x, y)\, dy\, dx = 1$.
3. $P((X, Y) \in A) = \iint_A f_{X,Y}(x, y)\, dx\, dy$, for every region $A \subset \mathbb{R}^2$, where $\mathbb{R}^2 \equiv \mathbb{R} \times \mathbb{R}$.

Result.
$$F_{X,Y}(x, y) \equiv P(X \leq x, Y \leq y) = \int_{-\infty}^{x} \int_{-\infty}^{y} f_{X,Y}(a, b)\, db\, da.$$

Result. If (X, Y) is a continuous random vector, then the marginal pdf of X is

$$f_X(x) = \int_{-\infty}^{\infty} f_{X,Y}(x, y)\, dy \text{ for every real number } x.$$

Analogously, the marginal pdf of Y is

$$f_Y(y) = \int_{-\infty}^{\infty} f_{X,Y}(x, y)\, dx \text{ for every real number } y.$$

Result.
$$f_{X,Y}(x, y) = \frac{\partial^2 F_{X,Y}(x, y)}{\partial x\, \partial y}.$$

Result. The marginal mean and variance of a continuous random variable X having marginal pdf f_X are the constants

$$E[X] \equiv \mu_X = \int_{-\infty}^{\infty} x f_X(x)\,dx = \int_{-\infty}^{\infty}\int_{-\infty}^{\infty} x f_{X,Y}(x,y)\,dy\,dx$$

and

$$\begin{aligned} V[X] \equiv \sigma_X^2 &= \int_{-\infty}^{\infty} (x-\mu_X)^2 f_X(x)\,dx \\ &= \int_{-\infty}^{\infty}\int_{-\infty}^{\infty} (x-\mu_X)^2 f_{X,Y}(x,y)\,dy\,dx. \end{aligned}$$

Analogous results hold for Y.

Comment. More generally, for any scalar function $g(X,Y)$ of random vector (X,Y) (where X and Y each can be either continuous, discrete or mixed)

$$E[g(X,Y)] \equiv -\int_{\mathcal{S}^-} P(g(X,Y) \le a)\,da + \int_{\mathcal{S}^+} P(g(X,Y) > a)\,da,$$

where $\mathcal{S}^-$ is the negative support of $g(X,Y)$ and $\mathcal{S}^+$ is the positive support of $g(X,Y)$. The specifics of these integrals depends on the functional form of $g(X,Y)$ and may involve use of the Jacobian method of change-of-variable from calculus.

Definition. For any differentiable scalar function $g(X,Y)$ of random vector (X,Y) (where X and Y each can be either continuous, discrete or mixed) that has $g(0,0) = g(x,0) = g(0,y) = 0$, the constant $\mathrm{E}\left[g\left(X,Y\right)\right] \equiv$

$$(-1)^2 \int_{-\infty}^{0} \int_{-\infty}^{0} \frac{\partial^2 g(a,b)}{\partial a\, \partial b}\Big|_{\substack{a=x\\b=y}} \mathrm{P}(X \le x, Y \le y)\, \mathrm{d}y\, \mathrm{d}x$$

$$+(1)^2 \int_{0}^{\infty} \int_{0}^{\infty} \frac{\partial^2 g(a,b)}{\partial a\, \partial b}\Big|_{\substack{a=x\\b=y}} \mathrm{P}(X > x, Y > y)\, \mathrm{d}y\, \mathrm{d}x$$

$$+(1)(-1) \int_{0}^{\infty} \int_{-\infty}^{0} \frac{\partial^2 g(a,b)}{\partial a\, \partial b}\Big|_{\substack{a=x\\b=y}} \mathrm{P}\left(X > x, Y \le y\right)\, \mathrm{d}y\, \mathrm{d}x$$

$$+(-1)(1) \int_{-\infty}^{0} \int_{0}^{\infty} \frac{\partial^2 g(a,b)}{\partial a\, \partial b}\Big|_{\substack{a=x\\b=y}} \mathrm{P}(X \le x, Y > y)\, \mathrm{d}y\, \mathrm{d}x$$

$$= \int_{-\infty}^{0} \int_{-\infty}^{0} \frac{\partial^2 g(a,b)}{\partial a\, \partial b}\Big|_{\substack{a=x\\b=y}} F_{X,Y}(x,y)\, \mathrm{d}y\, \mathrm{d}x$$

$$+ \int_{0}^{\infty} \int_{0}^{\infty} \frac{\partial^2 g(a,b)}{\partial a\, \partial b}\Big|_{\substack{a=x\\b=y}} \left(1 - (F_X(x) + F_Y(y)\right.$$
$$\left. - F_{X,Y}(x,y))\right)\, \mathrm{d}y\, \mathrm{d}x$$

$$- \int_{0}^{\infty} \int_{-\infty}^{0} \frac{\partial^2 g(a,b)}{\partial a\, \partial b}\Big|_{\substack{a=x\\b=y}} \left(F_Y(y) - F_{X,Y}(x,y)\right)\, \mathrm{d}y\, \mathrm{d}x$$

$$- \int_{-\infty}^{0} \int_{0}^{\infty} \frac{\partial^2 g(a,b)}{\partial a\, \partial b}\Big|_{\substack{a=x\\b=y}} \left(F_X(x) - F_{X,Y}(x,y)\right)\, \mathrm{d}y\, \mathrm{d}x.$$

Comment. When the joint probability density exists the result can be written

$$E[g(X,Y)] = \int_{-\infty}^{\infty} \int_{-\infty}^{\infty} g(x,y) f_{XY}(x,y) \, dx \, dy.$$

Comment. If X and/or Y is discrete, replace the corresponding integral with a summation.

Result. If (X, Y) is a continuous random vector having joint pdf $f_{X,Y}$, then the conditional pdf of Y given $X = x$, for $f_X(x) > 0$ and undefined otherwise, is

$$\begin{aligned} f_{Y|X}(y|x) &\equiv f_{Y|X=x}(y) \\ &\equiv \frac{f_{Y,X}(y,x)}{f_X(x)}. \end{aligned}$$

Comment. The conditional pdf of X given $Y = y$ is defined analogously.

Technical note. For continuous random variables, the conditioning appears to be on a zero-probability event and thus all conditional pdf's, conditional cdf's, and conditional expectations would not be defined (since we cannot divide by 0 and the probability of a continuous random variable being equal to a specific real-number value is always 0). This is a notational shorthand issue; in fact, we are conditioning on a continuous random variable taking on a value in some small neighborhood or interval. Specifically,

$$F_{Y|X}(y|x) \equiv P(Y \le y | X = x)$$

is shorthand notation for

$$F_{Y|X}(y|x) \equiv \lim_{h \to 0^+} P(Y \le y \mid x < X \le x + h)$$

-61-

$$= \lim_{h \to 0^+} \frac{P(-\infty < Y \le y \cap x < X \le x+h)}{P(x < X \le x+h)}$$

$$= \lim_{h \to 0^+} \frac{\int_x^{x+h} \left(\int_{-\infty}^y f_{X,Y}(a,b)\, db \right) da}{\int_x^{x+h} f_X(a)\, da}$$

$$= \lim_{h \to 0^+} \frac{\int_{-\infty}^y f_{X,Y}(x+h, b)\, db}{f_X(x+h)}$$

$$= \frac{\int_{-\infty}^y \lim_{h \to 0^+} f_{X,Y}(x+h, b)\, db}{\lim_{h \to 0^+} f_X(x+h)}$$

$$= \frac{\int_{-\infty}^y f_{X,Y}(x,b)\, db}{f_X(x)}$$

$$= \int_{-\infty}^y \frac{f_{X,Y}(x,b)}{f_X(x)}\, db$$

$$= \int_{-\infty}^y f_{Y|X}(b|x)\, db.$$

Comment.

- $P(A|B) = \frac{P(A \cap B)}{P(B)}$ for $P(B) > 0$, and
- $f_{X|Y}(x|y) = \frac{f_{X,Y}(x,y)}{f_Y(y)}$ for $f_Y(y) > 0$, however
- $F_{Y|X}(y|x) \ne \frac{F_{XY}(x,y)}{F_X(x)}$

Result. For every real number x at which $f_{Y|X}(\cdot|x)$ is defined,

- $f_{Y|X}(y|x) \ge 0$, for $y \in \mathbb{R}$,

– $\int_{-\infty}^{\infty} f_{Y|X}(y|x)\,dy = 1$, and

– $P(Y \in B | X = x) = \int_B f_{Y|X}(y|x)\,dy$,

for every interval (or set of intervals) $B \in \mathbb{R}$.

Result. The conditional mean and conditional variance of Y given $X = x$ are the constants

$$\begin{aligned} E[Y|X=x] &\equiv \mu_{Y|X=x} \\ &\equiv \int_{-\infty}^{\infty} y\, f_{Y|X}(y|x)\,dy \end{aligned}$$

and

$$\begin{aligned} V[Y|X=x] &\equiv \sigma^2_{Y|X=x} \\ &\equiv \int_{-\infty}^{\infty} (y - \mu_{Y|X=x})^2 f_{Y|X}(y|x)\,dy \\ &= E[Y^2|X=x] - \mu^2_{Y|X=x} \end{aligned}$$

Comment. The units of the constant $E[Y|X=x]$ are the units of Y, and the units of the constant $V[Y|X=x]$ are the units of Y^2. If x is not a specific real number but is an dummy variable then $E[Y|X=x]$ is a function of the dummy variable x as is $V[Y|X=x]$. In either case the units of $E[Y|X=x]$ and $V[Y|X=x]$ are the units of Y and Y^2, respectively.

Result. For every random vector (X, Y)

$$f_{X,Y}(x,y) = f_{X|Y}(x|y) f_Y(y) = f_{Y|X}(y|x) f_X(x),$$

for all $x, y \in \mathbb{R}$.

Definition. For every random vector (X, Y), X and Y are *independent* if and only if $f_{X,Y}(x,y) = f_X(x)f_Y(y)$ for all $x, y \in \mathbb{R}$.

Equivalently, X and Y are *independent* if and only if $F_{X,Y}(x,y) = F_X(x)F_Y(y)$ for all $x, y \in \mathbb{R}$.

(For completeness the above two results, stated in the section on joint discrete random variables, are repeated here).

Result. If X and Y are independent, then
$\mathrm{E}[g(X) \cdot h(Y)] = \mathrm{E}[g(X)] \cdot \mathrm{E}[h(Y)]$.

Result. For every random vector (X, Y) the following four statements are equivalent; that is, either none are true or all are true.

1. X and Y are independent.
2. $f_{Y|X}(y|x) = f_Y(y)$ for all $x, y \in \mathbb{R}$ with $f_X(x) > 0$.
3. $f_{X|Y}(x|y) = f_X(x)$ for all $x, y \in \mathbb{R}$ with $f_Y(y) > 0$.
4. $\mathrm{P}(X \in A, Y \in B) = \mathrm{P}(X \in A)\mathrm{P}(Y \in B)$ for all intervals (or set of intervals) A, and $B \subset \mathbb{R}$.

Definition. If (X, Y) is a random vector, then the *covariance* of X and Y is the constant

$$\begin{aligned}
\mathrm{Cov}[X, Y] &\equiv \mathrm{E}[(X - \mu_X)(Y - \mu_Y)] \\
&= \mathrm{E}[XY] - \mathrm{E}[X]\mathrm{E}[Y] \\
&= \mathrm{E}[XY] - \mu_X \mu_Y,
\end{aligned}$$

which is also denoted by σ_{XY}.

Comment. The units of $\mathrm{Cov}[X, Y]$ are the units of X times the units of Y.

Comment. $\mathrm{Cov}[X, Y] = \mathrm{Cov}[Y, X]$.

Comment. $\text{Cov}[X, X] = \text{V}[X]$.

Comment. Covariance is a measure of *linear* dependence.

Comment. If X and Y are independent then $\text{Cov}[X, X] = 0$.

Comment. If either $\text{V}[X]$ or $\text{V}[Y]$ or both are 0, then $\text{Cov}[X, Y] = 0$.

Comment. If $\text{Cov}[X, Y] = 0$, then possibly $\text{V}[X] > 0$, $\text{V}[Y] > 0$, and X and Y are dependent, but are not linearly dependent.

Definition. For random vector (X, Y) the *correlation* of X and Y is the constant

$$\text{Corr}[X, Y] \equiv \frac{\text{Cov}[X, Y]}{\sqrt{\text{V}[X]\text{V}[Y]}}.$$

Comment. $\text{Corr}[X, Y]$ is also denoted by $\rho_{X,Y}$ or, when not ambiguous, ρ.

Comment. $\rho_{X,Y}$ is not defined if either $\text{V}[X]$ or $\text{V}[Y]$ or both $= 0$.

Comment. $\rho_{X,Y}$ is a unitless constant.

Result. $-1 \leq \rho_{X,Y} \leq 1$.

Result. If $\rho_{X,Y}$ is defined and all observations of (X, Y) lie in a straight line, then $|\rho_{X,Y}| = 1$.

Comment. If all observations of (X, Y) lie in a straight line and if one (or both) of the random variables is (are) a constant; i.e., if one (or both) of the variances is (are) 0, then correlation is not defined and the covariance is 0..

Result. If $\rho_{X,Y}$ is defined and X and Y are independent, then $\sigma_{X,Y} = \rho_{X,Y} = 0$.

Comment. Covariance and correlation are measures of linear dependence. Zero correlation does not imply independence. Consider $Y = |X|$ for X uniformly distributed on $[-1, 1]$. Clearly X and Y are dependent (but are not linearly dependent), and yet $\text{Cov}[X, Y] = 0 = \text{Corr}[X, Y]$.

Nevertheless, informally the phrase "X and Y are correlated" sometimes is used to mean that "X and Y are dependent."

Result. For random vector (X, Y) having σ_x^2 and $\sigma_y^2 > 0$,

$$\begin{aligned}
\text{if } \text{E}[Y|X=x] &= a + bx, \\
\text{then } \text{E}[Y|X=x] &= \mu_Y + \rho \frac{\sigma_Y}{\sigma_X}(x - \mu_X), \\
\text{E}[Y|X] &= \mu_Y + \rho \frac{\sigma_Y}{\sigma_X}(X - \mu_X), \\
\text{and } \text{E}[\text{Var}[Y|X]] &= \sigma_Y^2(1 - \rho^2) \leq \sigma_Y^2
\end{aligned}$$

(even though $\text{Var}[Y|X=x] \neq \sigma_Y^2(1 - \rho^2)$).

i.e., if $\text{E}[Y|X=x]$ is a linear function in x, then we know the form of a and b.

Comment. $\text{Var}[Y|X=x]$ can be a function of x; however, $\text{E}[\text{Var}[Y|X]]$ is not a function of X, and is smaller than $\text{Var}[Y]$ if $\rho^2 > 0$.

Definition. Suppose that the continuous random vector (X, Y) has means (μ_X, μ_Y), positive variances (σ_X^2, σ_Y^2), and correlation ρ with $|\rho| < 1$. (i.e, zero variances and $|\rho| = 1$ are excluded.)

Then (X, Y) has the *bivariate normal distribution* if its pdf at every point (x, y) is

$$f_{X,Y}(x, y) = \frac{\exp\left(\frac{z_x^2 - 2\rho z_x z_y + z_y^2}{-2(1-\rho^2)}\right)}{2\pi \sigma_X \sigma_Y \sqrt{(1-\rho^2)}}$$

where $z_x = (x - \mu_X)/\sigma_X$ and $z_y = (y - \mu_Y)/\sigma_Y$ (i.e., the z-values of $X = x$ and $Y = y$).

Result. If (X, Y) is bivariate normal with parameters (constants) $\mu_X, \mu_Y, \sigma_X, \sigma_Y$, and ρ, then

1. the marginal distributions of X and Y are normal,
2. if $\rho = 0$ then X and Y are independent,
3. if $X = x$, the conditional distribution of Y is normal with
    - $\mu_{Y|X=x} = \mu_Y + \rho \sigma_Y z_x$ (a constant for every real number x),
    - $\sigma^2_{Y|X=x} = (1 - \rho^2)\sigma^2_Y$ (a constant for every real number x), and
4. if $Y = y$, the conditional distribution of X is normal with
    - $\mu_{X|Y=y} = \mu_X + \rho \sigma_X z_y$ (a constant for every real number y), and
    - $\sigma^2_{X|Y=y} = (1 - \rho^2)\sigma^2_X$ (a constant for every real number y).

Caution. If $X \sim N(\mu_X, \sigma^2_X)$ and $Y \sim N(\mu_Y, \sigma^2_Y)$ with positive variances and $-1 < \rho_{X,Y} < 1$, then (X, Y) is not necessarily bivariate normal.

Result. If Z_1 and Z_2 are two independent standard-normal random variables and if

$$X = \mu_X + \sigma_X Z_1$$
$$Y = \mu_Y + \sigma_Y \left(\rho_{X,Y} Z_1 + Z_2 \sqrt{1 - \rho^2_{X,Y}}\right),$$

then (X, Y) is a bivariate normal random vector with parameters (constants) $\mu_X, \mu_Y, \sigma_X, \sigma_Y$, and $\rho_{X,Y}$. (This result can be used to generate bivariate normal observations in a Monte Carlo simulation.)

Definition. Given random variables $X_1, X_2, \ldots, X_n$ and any constants $c_0, c_1, c_2, \ldots, c_n$, then the random variable

$$Y = c_0 + c_1 X_1 + c_2 X_2 + \cdots + c_n X_n$$

is a *linear combination* of the random variables $X_1, X_2, \ldots, X_n$ and is itself a random variable.

Result. The linear-combination random variable $Y = c_0 + c_1 X_1 + c_2 X_2 + \cdots + c_n X_n$ has mean and variance

$$\underbrace{\text{E}[Y]}_{\text{constant}} = \text{E}[c_0 + \sum_{i=1}^{n} c_i X_i] = \text{E}[c_0] + \sum_{i=1}^{n} \text{E}[c_i X_i]$$

$$= c_0 + \sum_{i=1}^{n} c_i \text{E}[X_i]$$

and

$$\underbrace{\text{V}[Y]}_{\text{constant}} = \sum_{i=1}^{n} \sum_{j=1}^{n} \text{Cov}[c_i X_i, c_j X_j]$$

$$= \sum_{i=1}^{n} \sum_{j=1}^{n} c_i c_j \, \text{Cov}[X_i X_j]$$

$$= \sum_{i=1}^{n} c_i^2 \, \text{V}[X_i] + 2 \sum_{i=1}^{n-1} \sum_{j=i+1}^{n} c_i c_j \, \text{Cov}[X_i X_j].$$

Comment. The additive constant, c_0, has no effect on $\text{V}[Y]$.

Corollary. If $X_1, X_2, \ldots, X_n$ are pairwise independent, then the linear-combination random variable $Y = c_0 + c_1 X_1 + c_2 X_2 + \cdots + c_n X_n$ has variance $\text{V}[Y] = \sum_{i=1}^{n} c_i^2 \, \text{V}[X_i]$ (a constant).

Definition. The *sample mean* of $X_1, X_2, \ldots, X_n$ is a random variable and is the linear combination
$$\bar{X} \equiv \bar{X}_n \equiv \frac{X_1 + X_2 + \cdots + X_n}{n}.$$

Notation. Common notation for the sample-mean random variable includes
$$\bar{X} \equiv \bar{X}_n.$$

Result. If $X_1, X_2, \ldots, X_n$ each has the same value for their means; i.e., $\mathrm{E}[X_i] \equiv \mathrm{E}[X] = \mu_X$ (a constant), for $i = 1, 2, \ldots, n$, then the mean of the sample-mean random variable, $\bar{X}$, is the constant
$$\mathrm{E}\left[\bar{X}\right] \equiv \mu_{\bar{X}} = \mu_X.$$

Comment. $\mathrm{E}\left[\bar{X}\right]$ is *not* a function of n, the sample size and is identical to the mean of the random variable X.

Result. If $X_1, X_2, \ldots, X_n$ are independent and each has the same value for their variances; i.e., $\mathrm{V}[X_i] = \mathrm{V}[X] = \sigma_X^2$ (a constant), for $i = 1, 2, \ldots, n$, then the variance of the random variable $\bar{X}$ is the constant
$$\mathrm{V}\left[\bar{X}\right] \equiv \sigma_{\bar{X}}^2 = \sigma_X^2/n.$$

Comment. $\mathrm{V}\left[\bar{X}\right]$ *is* a function of n, and decreases to 0 as n increases to ∞.

Comment. Even though the random variable X_i, for $i = 1, 2, \ldots, n$, and the associated sample-mean random variable, $\bar{X}$, have the same mean, they have different variances when $n > 1$.

Result. *Reproductive Property of the Normal Distribution.* If $X_1, X_2, \ldots, X_n$ are independent normally distributed random variables, then the linear-combination random variable $Y = c_0 + c_1 X_1 + c_2 X_2 + \cdots + c_n X_n$ is a normally distributed random variable; i.e.,

$$Y \sim \mathrm{N}\left(\underbrace{c_0 + \sum_{i=1}^{n} c_i \mu}_{\mu_Y},\ \underbrace{\sum_{i=1}^{n} c_i^2 \mathrm{V}[X_i] + 2 \sum_{i=1}^{n-1} \sum_{j=i+1}^{n} c_i c_j \mathrm{Cov}[X_i X_j]}_{\sigma_Y^2} \right),$$

or more succinctly

$$Y \sim \mathrm{N}\left(\mu_Y, \sigma_Y^2\right),$$

where $\mu_Y \equiv \mathrm{E}[Y]$ and $\sigma_Y^2 \equiv \mathrm{V}[Y]$ are constants.

Corollary. If $X_1, X_2, \ldots, X_n$ are mutually independent normal random variables with common mean μ_X (constant) and common variance σ_X^2 (constant), then the sample mean random variable, $\bar{X}$, is

$$\bar{X} \sim \mathrm{N}(\mu_X, \sigma_X^2/n).$$

Result. *Central Limit Theorem.* For "large" random samples, the sample mean is approximately normally distributed with mean μ_X and variance σ_X^2, regardless of the distributions of $X_1, X_2, \ldots, X_n$. (That is, when n is "large" the random variable $\bar{X}$ can be validly assumed to be normally distributed with mean μ_X (constant) and variance σ_X^2/n (constant).)

More precisely: Let $\bar{X}_n$ be the sample-mean random variable from n independent observations having common mean, μ_X, and common variance, σ_X^2, and let

$$Z \equiv \lim_{n \to \infty} \frac{\bar{X}_n - \mu_X}{\sigma_X/\sqrt{n}}.$$

Then $Z \sim \mathrm{N}(0,1)$.

Descriptive Statistics

Comment. A *sample* of *data*, denoted here by $x_1, x_2, \ldots, x_n$, can be summarized (that is, described) numerically or graphically.

Definition. A *statistic*, $\widehat{\Theta}$, is a function of a sample $X_1, X_2, \ldots, X_n$.

Comment. Because a statistic is a function of a sample $X_1, X_2, \ldots, X_n$, then a statistic is a random variable.

Notation. $\hat{\theta}$ is the numerical value of the statistic $\widehat{\Theta}$.

Comment. Most of the statistics studied in elementary courses are functions of *random* samples; i.e., random variables $X_1, X_2, \ldots, X_n$ are independent and identically distributed (iid); i.e., each of the independent random variables, X_i, has the same distribution as X, where X has pdf (pmf) and cdf, $f_X(x)$ and $F_X(x)$, respectively.

Definition. The iid random variables $X_1, X_2, \ldots, X_n$ are called *random observations* and compose the random sample.

Definition. The real-number numerical values $x_1, x_2, \ldots, x_n$ are called *observations* of the random sample $X_1, X_2, \ldots, X_n$.

Commonly used statistics include:

- sample-average numerical value, $\bar{x}$, (the center of gravity of the data, a data measure of location):

$$\bar{x} = n^{-1} \sum_{i=1}^{n} x_i$$

Comment. The sample-average numerical value $\bar{x}$ is an observation of the random-variable sample mean $\bar{X}$.

- sample variance (the moment of inertia of the data, a measure of dispersion):

$$s^2 = \frac{\sum_{i=1}^{n}(x_i - \bar{x})^2}{n-1} = \frac{\sum_{i=1}^{n} x_i^2 - n\bar{x}^2}{n-1}$$

Comment. The sample-variance numerical value s^2 is an observation of the random-variable sample variance S^2.

- sample standard deviation of the data (an alternative measure of dispersion):

$$s = +\sqrt{s^2}$$

Comment. The sample standard deviation numerical value s is an observation of the random-variable sample variance S.

- sample range of the data (an alternative measure of dispersion). Often $4s \leq r \leq 6s$.

$$\begin{aligned} r &= \max\{x_1, x_2, \ldots, x_n\} - \min\{x_1, x_2, \ldots, x_n\} \\ &= x_{(n)} - x_{(1)}. \end{aligned}$$

Comment. The sample range of the data is an observation of the sample-range random variable R.

- $100k$ sample percentiles for $k = .01, .02, \ldots, .99$ of the data.
  (alternative measures of location):
  $p_i = $ the data value greater than (approximately) $100k\%$ of the data.

Comment. The $100k$ sample percentiles of the data are observations of the $100k$ sample percentile random variables P_i.

- first, second, and third sample quartiles of the data (alternative measures of location):

$$q_1 \equiv p_{25}$$
$$q_2 \equiv p_{50} \, (\equiv \text{median of the data})$$
$$q_3 \equiv p_{75}$$

Comment. The first, second, and third sample quartiles of the data are observations of the first, second, and third sample quartile random variables Q_i.

- sample median of the data (an alternative measure of location):

$$\tilde{m} \equiv q_2 \equiv p_{50}$$

Comment. The sample median of the data is an observation of the sample median random variable $\widetilde{M}$.

- sample inter-quartile range of the data (an alternative measure of dispersion):

$$iqr \equiv q_3 - q_1 = p_{75} - p_{25}$$

Comment. The sample inter-quartile range of the data is an observation of the sample inter-quartile range random variable IQR.

- sample mode of the data (the most commonly occurring data value, an alternative measure of the location).

Result. For constants a and b and observations $x_1, x_2, \ldots, x_3$, consider the set of *coded data* $y_i = a + bx_i$ for $i = 1, 2, \ldots, n$. Then

- *location* measures are multiplied by b and then increased by a,
- *dispersion* measures (except the sample variance) are multiplied by $|b|$, and
- the *sample variance* is multiplied by b^2.

Graphical summaries of sets of observations include dot plots, histograms, cumulative distribution plots, stem-and-leaf diagrams, and box plots. Data values plotted against time is a time-series plot; appending a stem-and-leaf plot to a time-series plot yields a digidot.

- *Frequency* is the number of observations (numerical data values) satisfying a specified condition, such as lying in a particular interval. *Relative frequency* is the frequency divided by n, the fraction of the observations satisfying the condition.
- A *histogram* is a bar graph showing observation-value frequency for several adjacent equal-length intervals. (The intervals are sometimes called *cells* or *bins*.)
- A *cumulative distribution plot* is analogous to a histogram, but each bar shows *cumulative frequency*, the number of observation values in the bin or to the left of the bin. (Alternatively, each bar can show *cumulative relative frequency*.)
- An *empirical cdf plot* re-scales the cumulative distribution plot, replacing frequency with relative frequency.
  Plot $\left(x_{(i)}, \frac{i}{n+1}\right)$, for $i = 1, 2, 3, \ldots, n$, recalling that $x_{(i)}$ is the ith ordered observation.

Parameter Estimation

Comment. *Inferential Statistics* is the study of data drawn from a system to infer conclusions about the system, a deeper topic than descriptive statistics.

Definition. A *population* is the set of all possible observations of the relevant system. (For example, the students in a class.)

Definition. A constant θ is a *population parameter* if it is a characteristic of the population. (For example, a class's average gpa.)

Definition. A *sample* is a subset selected from the population. (For example, the students in the front row of a class.)

Definition. The random vector $(X_1, X_2, \ldots, X_n)$ is *independent and identically distributed*, often abbreviated *iid*, if

- the X_i's are mutually independent and
- every X_i has the same probability distribution, say with cdf F_X.

Definition. The random vector $(X_1, X_2, \ldots, X_n)$ is a *random sample* (of size n) if it is iid.

(For now, every random sample is iid. Later, the definition will be generalized to discuss more-sophisticated sampling ideas. At that time this simplest kind of sample will become an "iid random sample" or a "simple random sample".)

Definition. A *statistic* $\hat{\Theta}$ is (a random variable that is) a function $h(\cdot)$ of a random sample prior to choosing the sample.
$$\hat{\Theta} = h(X_1, X_2, \ldots, X_n).$$
(For example, the average gpa of the students in the sample prior to choosing the sample.)

Definition. A statistic $\hat{\Theta}$ (a random variable) is a *point estimator* of the population parameter θ (a constant) if its purpose is to guess the value of θ.

Definition. A *point estimate* $\hat{\theta}$ is a single (real-number valued) observation of the random variable $\hat{\Theta}$. ("$\hat{\Theta} = \hat{\theta}$" is an event in the same sense that "$X = x$" is an event.)

Definition. A sequence of point estimators $\hat{\Theta}_n$ (a sequence of random variables) is *consistent* if $\lim_{n \to \infty} P(|\hat{\Theta}_n - \theta| < \epsilon) = 1$ for every positive constant ϵ. (Usually n is sample size. A consistent estimator, then, is guaranteed to be arbitrarily close to the constant θ for large sample sizes.)

Definition. The *bias* of the (random variable) point estimator $\hat{\Theta}$ is the constant
$$\text{Bias}\left[\hat{\Theta}, \theta\right] = \text{E}[\hat{\Theta}] - \theta.$$

Definition. The point estimator $\hat{\Theta}$ (a random variable) is an *unbiased* estimator of θ if $\text{E}[\hat{\Theta}] = \theta$, a constant.

Definition. The *standard error* of a point estimator $\hat{\Theta}$ (a random variable) is its standard deviation, the constant
$\sigma_{\hat{\Theta}} \equiv \sqrt{\text{V}[\hat{\Theta}]}.$
(For example, $\sigma_{\bar{X}} = \sigma_X/\sqrt{n}$.)

Definition. The *mean squared error* (*MSE*) of a point estimator $\hat{\Theta}$ of the parameter θ is the constant

$$\text{MSE}[\hat{\Theta}, \theta] \equiv \text{E}[(\hat{\Theta} - \theta)^2].$$

Result. $\text{MSE}[\hat{\Theta}, \theta] = \text{Bias}^2[\hat{\Theta}, \theta] + \text{V}[\hat{\Theta}]$.
(Here $\text{Bias}^2[A, b] \equiv (\text{Bias}[A, b])^2$.)

Definition. The *root mean squared error* (*RMSE*) (a constant) is the non-negative square root of the MSE.

Comments.

- The concept of variance is generalized by MSE, which is useful when the point estimator is biased. RMSE is analogous to standard deviation.
- The units of the MSE are the same as the square of the units of $\hat{\Theta}$.
- The units of the RMSE are the same as the units of $\hat{\Theta}$.
- For most commonly used point estimators, squared bias goes to zero faster than variance as n increases, so asymptotically $\text{MSE}[\hat{\Theta}, \theta]/\text{V}[\hat{\Theta}] = 1$.
- Some biased point estimators are good in the sense of having a small MSE.

Table III – Point Estimators

Distribution Parameter (constant)	Point Estimator (random variable)	Sampling Distribution Mean (constant)
θ	$\widehat{\Theta}$	$E[\widehat{\Theta}]$
$p \equiv P(A)$	$\widehat{P} \equiv \text{"\# of successes"}/n$	p
$\mu \equiv E[X]$	$\bar{X} \equiv \sum_{i=1}^{n} X_i/n$	μ
$\mu_1 - \mu_2$	$\bar{X}_1 - \bar{X}_2$	$\mu_1 - \mu_2$
$\sigma^2 \equiv V[X]$	$S^2 \equiv \sum_{i=1}^{n}(X_i - \bar{X})^2/(n-1)$	σ^2

Definition. The *minimum-variance unbiased estimator* (*MVUE*) is the unbiased estimator of θ having the smallest variance.

(More precisely, for the iid random vector $(X_1, X_2, \ldots, X_n)$ drawn from a particular distribution having parameter (constant) θ, consider all functions $h(\cdot)$ for which $E[h(X_1, X_2, \ldots, X_n)] = \theta$. The MVUE of θ is the point estimator defined by the function $h(\cdot)$ that minimizes $V[h(X_1, X_2, \ldots, X_n)]$).

Result. If the random vector $X_1, X_2, \ldots, X_n$ is a random sample from a distribution with (constant) mean μ and (constant) variance σ^2, then the sample-mean random variable $\bar{X}$ is the MVUE for μ.

(We already knew that $\bar{X}$ is unbiased for μ, and that $V[\bar{X}] = \sigma^2/n$. The new point is that the functional form $h(\cdot)$ of the MVUE is that of a sample average.)

iid Sampling Distribution Variance (constant)	Standard-Error Estimator (random variable)
$V[\widehat{\Theta}] \equiv [\text{ste}(\widehat{\Theta})]^2$	$\widehat{\text{ste}}(\widehat{\Theta}) \equiv \widehat{\sigma}_{\widehat{\Theta}}$
$p(1-p)/n$	$\sqrt{\widehat{P}(1-\widehat{P})/(n-1)}$
σ^2/n	$S/\sqrt{n}$
$\sigma_1^2/n_1 + \sigma_2^2/n_2$	$\sqrt{S_1^2/n_1 + S_2^2/n_2}$
$\sigma^4 n^{-1}(\alpha_4 - (n-3)/(n-1))^*$	difficult

* Here the quantity α_4 is the fourth standardized moment or *kurtosis*; i.e., $\alpha_4 \equiv E[(X-\mu)^4]/\sigma^4$.

Definition. The distribution of a point estimator (random variable) $\widehat{\Theta}$ is its *sampling distribution*.

- The quality of a (random variable) point estimator is determined by its sampling distribution.
- The sampling distribution (and therefore the (constant) bias and (constant) standard error) of random variable $\widehat{\Theta}$ depends upon the sample size n, the distribution F_X, independence, and the function $h(\cdot)$.
- For all (random variable) point estimators in a first course, and for almost all (random variable) point estimators in general, as n becomes large
  - the (constant) bias goes to zero (at a rate inversely proportional to n),

- the (constant) standard error goes to zero (at a rate inversely proportional to $\sqrt{n}$), and
- the sampling distribution becomes normal.

Definition. The *estimated standard error*, a random variable, is $\widehat{\sigma}_{\widehat{\Theta}}$ (or $\widehat{\text{ste}}(\widehat{\Theta})$), is a point estimate of the standard error, $\text{ste}(\widehat{\Theta})$, (a constant).

- A common notational convenience, illustrated here, is to denote an estimator (a random variable) by placing a "hat" over a the quantity being estimated (a constant).
- The reason to estimate a standard error $\sigma_{\widehat{\Theta}}$ is to evaluate the quality $\widehat{\Theta}$.
- The reason to estimate θ is to make an inference about the system of interest.
- Many point estimators (random variables) are created by estimating unknown constants. (For example, $\hat{\sigma}_{\bar{X}} = S/\sqrt{n}$, where S is the sample standard deviation (a random variable).)
- Definition. The *bootstrap* is an alternative method of estimating the constant $\sigma_{\hat{\theta}}$, especially when the function $h(\cdot)$ is complicated.
- Definition: *Bootstrapping* is a general, but computationally intensive, method to estimate the standard error of any point estimator, say $\widehat{\Theta}$. Let $x_1, x_2, ..., x_n$ denote the sample from which the point estimate $\hat{\theta}$ was computed. Consider all n^n possible size-n samples composed of the values, selected with replacement, from the observed sample; for the ith possible sample, there is a corresponding point-estimate value, say $\hat{\theta}_i$. An estimate of the standard error is the standard

deviation of the n^n possible values:

$$\widehat{\text{ste}}\left(\hat{\Theta}\right) = \left[\frac{\sum_{i=1}^{n^n} \hat{\theta}_i^2 - n^n \bar{\theta}^2}{n^n - 1}\right]^{1/2},$$

where $\bar{\theta}$ denotes the average of the n^n point-estimate values; i.e.,

$$\bar{\theta} \equiv n^{-n} \sum_{i=1}^{n^n} \hat{\theta}_i.$$

When $n > 4$, computation can be saved with a Monte Carlo simulation approximation. Independently choose m of the n^n possible samples, where m is much smaller than n^n, say $m = 200$. If $\hat{\theta}_i$ now corresponds to the ith randomly chosen sample, the bootstrap standard-error estimate is

$$\widetilde{\text{ste}}\left(\hat{\Theta}\right) = \left[\frac{\sum_{i=1}^{m} \hat{\theta}_i^2 - m\tilde{\theta}^2}{m - 1}\right]^{1/2},$$

where $\tilde{\theta}$ denotes the average of the m Monte Carlo point-estimate values; i.e.,

$$\tilde{\theta} \equiv m^{-1} \sum_{i=1}^{m} \hat{\theta}_i.$$

- Comment: Think about why the bootstrap estimate is not divided by $\sqrt{m}$.

Distribution fitting. Choosing values (constants) of the distribution parameters to obtain the desired distributional properties. Two classical methods are i) The Method of Moments and ii) Maximum Likelihood Estimation.

- Method of Moments (MOM). Fitting a k-parameter distribution to a real-world context by matching the values of the first k distribution moments to the corresponding k sample moments (constants).
- Maximum Likelihood Estimation (MLE).
    - Definition. The *likelihood* $L(\cdot)$ of a set of observations (the random sample) $x_1, x_2, \ldots, x_n$ is $f_{X_1, X_2, \ldots, X_n}(x_1, x_2, \ldots, x_n)$, the n-dimensional joint pmf (if discrete) or joint pdf (if continuous) evaluated at the observed values $x_1, x_2, \ldots, x_n$.
    - Definition. The *likelihood function* $L(\theta)$ of an observed random sample $x_1, x_2, \ldots, x_n$ is $f_{X_1, X_2, \ldots, X_n}(x_1, x_2, \ldots, x_n; \theta)$, where θ is a distribution parameter (a constant).
    - Result. Assume that $x_1, x_2, \ldots, x_n$ are the observed values of a random sample, $X_1, X_2, \ldots, X_n$, where the pmf or pdf is $f(x; \theta)$. Then the sample's likelihood function is $L(\theta) =$

    $$f(x_1;\theta) f(x_2;\theta) \cdots f(x_n;\theta) = \prod_{i=1}^{n} f(x_i;\theta),$$

    (The observed sample is known, so $L(\theta)$ is a function of only the unknown parameter θ. The analyst must assume that the observations x_i are observed values of the random variable X and that X has a probability distribution from a family of distributions parameterized by θ (a constant); for example, the normal family with $\theta = (\mu, \sigma^2)$.)

- Definition. The *maximum likelihood estimator* (MLE) of θ is the (feasible) value of θ that maximizes $L(\theta)$.
- Result. The value of θ that maximizes $L(\cdot)$ also maximizes any continuous monotonic function of $L(\cdot)$. (In particular, $L(\cdot)$ and $\ln L(\cdot)$ are both maximized by the same value of θ, a useful result because often $\ln L(\cdot)$ is more tractable, especially if maximization is accomplished by setting the first derivative to zero.)
- Result. Except in unusual situations, MLEs have these large-sample properties:
    - approximately unbiased,
    - nearly minimum variance,
    - approximately normally distributed, and
    - consistent.
- Result *MLE Invariance*. Let the random variables $\widehat{\Theta}_1, \widehat{\Theta}_2, \ldots, \widehat{\Theta}_k$ be the MLEs of the parameters $\theta_1, \theta_2, \ldots, \theta_k$ (constants). The MLE of any function $h(\theta_1, \theta_2, \ldots, \theta_k)$ is the same function $h(\widehat{\Theta}_1, \widehat{\Theta}_2, \ldots, \widehat{\Theta}_k)$ of the estimators $\widehat{\Theta}_1, \widehat{\Theta}_2, \ldots, \widehat{\Theta}_k$. (For example, if the MLE of σ^2 is $\hat{\sigma}^2$, then the MLE of σ is $\hat{\sigma}$.)

The Greek Alphabet

name	lower case	upper case
alpha	α	A
beta	β	B
gamma	γ	Γ
delta	δ	Δ
epsilon	ϵ	E
zeta	ζ	Z
eta	η	H
theta	θ	Θ
iota	ι	I
kappa	κ	K
lambda	λ	Λ
mu	μ	M
nu	ν	N
xi	ξ	Ξ
omicron	o	O
pi	π	Π
rho	ρ	P
sigma	σ	Σ
tau	τ	T
upsilon	υ	Y
phi	ϕ	Φ
chi	χ	X
psi	ψ	Ψ
omega	ω	Ω

A Few Mathematical Results

Fubini's Theorem. Interchanging the order of summation.
$$\sum_{n=0}^{\infty}\sum_{k=0}^{n} a(n,k) = \sum_{k=0}^{\infty}\sum_{n=k}^{\infty} a(n,k),$$
when either the left-hand side or the right-hand side converges.

Sum of the positive integers.
$$\sum_{\substack{i=0\\i=1}}^{m} i = \frac{m(m+1)}{2}.$$

Geometric finite sum.
$$\sum_{i=0}^{m} a^i = \frac{1-a^{m+1}}{1-a}, \text{ for } |a| \neq 1.$$

Geometric infinite sum.
$$\sum_{i=0}^{\infty} a^i = (1-a)^{-1}, \text{ for } |a| < 1.$$

Matrix geometric infinite sum.
$$\sum_{i=0}^{\infty} A^i = (I-A)^{-1}, \text{ for } \det|A| \neq 0.$$

MacLaurin Series.
$$f(x) = \sum_{i=0}^{\infty} \frac{f^{(i)}(0)x^i}{i!},$$
where $f^{(i)}(0) \equiv \frac{d^i f(a)}{da^i}|_{a=0}$, for all x in some interval about $x=0$, for all functions $f(x)$ that have continuous uniformly bounded derivatives of all orders in that interval.

Power-series expansion of e.

$$e^x = \sum_{i=0}^{\infty} \frac{x^i}{i!}, \text{ for all } x.$$

Binomial expansion. For positive integers n,

$$(a+b)^n = \sum_{i=0}^{n} \binom{n}{i} a^{n-i} b^i.$$

The special case of the binomial, $a = 1$ and $b = 1$, yields

$$2^n = \sum_{i=0}^{n} \binom{n}{i}.$$

Normal constant.

$$\int_{-\infty}^{\infty} e^{\frac{1}{2}x^2} \, dx = \sqrt{2\pi}.$$

Exponential integral.

$$\int_a^b e^{-\alpha x} \, dx = -\frac{1}{\alpha} e^{-\alpha x} \Big|_a^b = -\frac{1}{\alpha}(e^{-\alpha b} - e^{-\alpha a})$$

Integration by parts.

$$\int u \, dv = uv - \int v \, du.$$

A Useful Differential Equation.

$$\frac{\mathrm{d}y(x)}{\mathrm{d}x} + a(x)\, y(x) = f(x).$$

The general solution to this differential equation is

$$y(x) = \exp\left(-\int a(x)\,\mathrm{d}x\right) \int \exp\left(\int a(x)\,\mathrm{d}x\right) f(x)\,\mathrm{d}x$$
$$+ c \cdot \exp\left(-\int a(x)\,\mathrm{d}x\right).$$

The particular solution is found by evaluating the constant c using a boundary condition.

Made in the USA
Lexington, KY
28 August 2012